D1356125

SEEING
PHYSICS

2,600 YEARS OF DISCOVERY: FROM THALES TO HIGGS

SEEING PHYSICS

2,600 YEARS OF DISCOVERY: FROM THALES TO HIGGS

PROFESSOR
DON LEMONS
with drawings by Jesse Graber

ANDRE
DEUTSCH

Published in 2018 by André Deutsch
An imprint of the Carlton Publishing Group
20 Mortimer Street
London W1T 3JW

10 9 8 7 6 5 4 3 2 1

A CIP catalogue record for this book is available from the British Library.

ISBN 978 0 233 00538 6

Printed in Dubai

Contents

Nineteenth Century

Twentieth Century and Beyond

Preface

Galileo's claim that mathematics is the language of science applies to no science more than to physics. But mathematical description requires great effort. How is that effort begun? The answer, more often than not, is with a drawing of an as yet wordless, pre-mathematical picture of reality. To draw is to see the world in a particular way and to inform the self with an understanding of the world. Drawing the important elements of physical reality diminishes the psychological difficulty of articulating that reality in language of any kind. Subsequent progress allows one to refine an initially crude drawing.

Carefully constructed drawings play a large role in teaching and in learning physics. Routinely I require my own students, when beginning to analyse a physical situation, to produce a drawing that represents important physical elements placed in right relation. Many of my students are visual learners and require little inducement, but the task benefits all. The drawing or diagram produced, sometimes called "the cartoon approximation", guides the whole process of investigation.

Drawings are a humble but effective tool of the physicist's craft and part of the tradition of physics that is passed on from colleague to colleague and from teacher to student. Certain drawings achieve relative fame and permanence on the pages of research journals, monographs, and textbooks. Many more exist only briefly on chalkboards or scraps of paper.

Drawings that jump-start a physical inquiry and encapsulate its results require neither rich detail nor realistic perspective – only simplicity and clarity. A good physics drawing is in many ways like a good epigram: spare; and, once composed, its elements cannot be subtracted from, added to, or rearranged without diminishing the composition. Also, like a good epigram, a good drawing is worth committing to memory.

This volume contains fifty-one exemplary drawings from twenty-six centuries of physics discovery arranged in chronological order. Each drawing presents a single idea. Most of the drawings appear frequently in other physics books, and all have survived the test of my own teaching and learning. Each drawing anchors an episode of the story I tell. An essay reviews the physics and places it in historical context.

When I started this project I was not sure that physics could be presented in any depth or breadth in this way, but I wanted to try. I am, after all, a theoretician who uses mathematics in research and a teacher whose duty it has been to find the mathematics most appropriate for his students. But to abstract the essential physics from a complicated situation and to represent that physics in a drawing is also close to the work of a theoretical physicist and teacher. I was delighted to learn that the word *theory* is related, through its Greek roots, to the word for *seeing*. And, of course, the phrase *I see* often means *I understand.*

Here is the result – admittedly an episodic and incomplete account. I am satisfied, but an unbiased evaluation necessarily falls to others. This book is for readers interested in the world in which they live but who, for various reasons, know little mathematics or physics. My hope is that *Seeing Physics* will, by appealing to their visual sense, help these readers say, "Now I see, and now I understand."

Dedications and Acknowledgements

My father, the late Reverend Wishard F. Lemons, would have been pleased that, finally, I had written a book he could read. Another dear one who has passed on, Anthony Gythiel, friend, literary scholar, and medievalist, encouraged me to include essays on medieval physics. Memory Eternal, Wishard and Tony! To you and to my young grandsons, Abel and Emil, I dedicate this book.

I also want to acknowledge friends and family members Galen Gisler, Christina Gore, Clark Lemons, Rick Shanahan, and David Watkins, whose reading and critique of the text helped improve it. Dan Umansky's class on Renaissance history inspired the essay on Leonardo. Tom Delillo's class on mathematical modelling inspired the essay on global warming. Jeremy Bernstein answered my questions on Einstein. Terrence Figy, Holger Meyer, and Nick Solomey advised me on the Higgs boson. The seminars in which I participated at St. John's College in Santa Fe, New Mexico, encouraged me to study the original sources upon which many of these essays are based. The Natural Science Seminar at Bethel College of North Newton, Kansas, responded helpfully to a presentation of the text.

I am long overdue in acknowledging Trevor Lipscombe for editing my earlier books and for giving me good advice on this one. John Thornton, my agent, placed *Seeing Physics* with Carlton Books. Thanks are due to Trevor and John and the helpful staff at Carlton Books. And I don't believe I could have found a better illustrator for *Seeing Physics* than Jesse Graber. My teachers and mentors over the decades, Robert Armstrong, Harold Daw, Peter Gary, Michael E. Jones, Robert Romer, and Dan Winske, helped prepare me, in various ways, for the task of writing this book. With pleasure, I offer this text to Bill Peter, physicist and longtime friend.

Finally, with much feeling I acknowledge Allison, my beloved wife of many years, for her support of my writing habit, for her critical reading, and for her many suggestions that have strengthened *Seeing Physics*.

Antiquity

1. Triangulation (600 BCE)

Figure 1

When a surveyor cannot measure a certain distance directly, say the width of a river or the height of a tree, either by counting paces or by laying out lengths of a standard measure, he can use the properties of triangles to determine the distance. This idea, which goes back to Thales of Miletus (624–565 BCE), is one of the first in the history of physics and mathematics.

Miletus was, in the 6th century BCE, a Greek port on an island off the west coast of Asia Minor, now modern Turkey, and Thales was an early philosopher or "lover of wisdom". Thales travelled far from Miletus in his search for wisdom – to Babylon and across the eastern Mediterranean to Egypt. Egypt, even in the 6th century

BCE, was known for its ancient civilization. After all, the great pyramids were built in approximately 2500 BCE. What Thales found in Egypt, if not wisdom, was the practical knowledge of local Egyptian land measurers or geometers who were skilled at measuring the position, size, and shape of agricultural plots, presumably that they might not be lost or confused with neighbouring plots after an episode of Nile flooding.

How did Thales convert the practical knowledge of the Egyptian land measurers to the universally applicable principles of geometrical surveying we now call triangulation? He may have been helped with a diagram such as that of the tree and the rod. When an upright rod casts a shadow equal in length to its height, one can expect that every other upright object will also cast a shadow equal in length to its height. Thus, when the height of the rod is equal to the length of its shadow, the unknown height of the tree is equal to the length of its easily measured shadow.

Such an inference requires that different rays of sunlight are all straight and parallel to one another. This supposition also allows us to use a rod and its shadow to determine the height of the tree at any time of day since the sides of all similarly shaped triangles stand in the same relation to one another. For whenever an upright object casts a shadow, a right-angled triangle is formed out of the object, its shadow, and a line connecting the top of the object to the top of its shadow. Thus, the ratio of the height of the object to its shadow is the same for all upright objects at any one time and location.

Figure 2 shows two such triangles: one formed by a taller object and its shadow and the other formed by a shorter one and its shadow. Both shadows are shorter than the objects are tall. Since the two triangles have the same shape, the ratios of their heights to the length of their shadows, H/H' and h/h', must be the same, that is $H/H'=h/h'$. Therefore, the presumably unknown height of the taller object can be expressed in terms of the directly measurable quantities H', h' and h', in the formula $H=H'h/h'$. Note that one need not measure an angle in order to use this method.

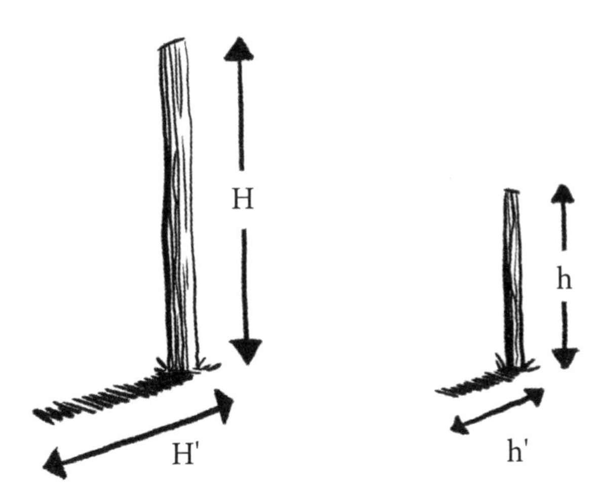

Figure 2

Although we have no record in Thales' own words, secondary sources credit him with measuring the height of the great pyramid at Giza and with determining the distance from the shore to ships at sea – possibly with methods similar to those described here. Today our smart phones and global positioning satellites also exploit the properties of similar triangles.

Thales is also said to have discovered how to inscribe a right-angled triangle within a circle – a discovery in gratitude for which he sacrificed an ox. He speculated that the principle, or source, of all things is water; he shifted the course of a river; and he correctly predicted, within a year, the occurrence of a relatively rare event: the Moon completely obscuring the Sun in a solar eclipse. For these feats of skill and wisdom Thales was honoured as one of the seven wise men of antiquity.

Thales, unlike those who seek only practical mastery, sought universal truths among the diversity of particular facts. He was a philosopher. And for his application of mathematical truths to the natural world he could also be called the first physicist.

2. Pythagorean Monochord (500 BCE)

Figure 3

One of the simplest musical instruments imaginable, the Pythagorean monochord, is a single stretched string fixed at each end. When plucked the string vibrates and produces a tone of a particular pitch. Longer and heavier strings produce lower tones just as do longer and larger wind and percussion instruments. These facts must have been known before the time Pythagoras flourished around 525 BCE. After all, musical instruments with several strings of different lengths, such as the ood and the lyre, are depicted on Greek vases that date from the seventh century BCE. But Pythagoras, or one of his followers, may have been the first to quantify the relationship between the length of a string and the tone it produces.

Around 525 BCE Pythagoras emigrated from his native island of Samos, near the west coast of Asia Minor in the Aegean Sea, to the Dorian Greek colony of Croton (modern day Crotone) near the ball of the boot that the coast of southern Italy outlines. There he founded a brotherhood of scholars who practised a discipline whose object was to care for and purify the soul. The brotherhood also aspired to be the beneficent if austere political leaders of Croton. Around 450 BCE the original brotherhood was overthrown and broken up, but mystics and scholars called Pythagoreans were prominent for at least another hundred years. Some of them were mathematically talented investigators who attributed their own discoveries to their leader Pythagoras.

Figure 3 depicts a Pythagorean monochord. We now know that the tone produced by a monochord is determined by the dominant frequency of its vibration and that this frequency is in turn determined by the length of the string (the longer the string, the lower the pitch) and by the speed of a disturbance on the string (the faster the disturbance, the higher the pitch). We also know that strings of lesser density and strings under greater tension produce more quickly travelling disturbances and, therefore, produce higher frequencies and higher pitches. Yet in no way does this knowledge reduce the mystery of the relation, discovered by the Pythagoreans, between whole numbers and pleasing sounds.

Imagine two such monochords with strings of identical composition and tension but unequal in length. When simultaneously plucked or struck the two strings produce two different tones. This humble arrangement allowed the Pythagoreans to make a discovery that filled them, and today fills us, with wonder. When one monochord is twice as long as a similarly constructed monochord, or, more generally, when the lengths of the monochord strings are to one another as two small whole numbers, such as 2 to 1, 3 to 2, or 4 to 3, then when plucked at the same time they produce, respectively, the pleasing sound of an octave, a perfect fifth, or a perfect fourth. Otherwise the tones are not so pleasing, but rather inharmonious or discordant.

That the small whole numbers 1, 2, 3, and 4 should correspond to pleasing sounds became for the Pythagoreans emblematic of the numeric nature of our world. According to the Pythagoreans both the form and the substance of the world are composed of whole numbers. Thus, for instance, the soul is a numerical harmony of the parts of the body. Even particular qualities such as "maleness" and "femaleness" are associated with numbers – in this case odd and even numbers, respectively. Today we find such ideas both vague and arbitrary. But the idea of finding common ratios and numerical forms in various phenomena is consistent with modern physics.

The Pythagorean monochord is the most basic of stringed

instruments – hardly a musical instrument at all. But it demonstrates the principle behind the way violins, harps, and all stringed instruments make pleasing sounds. Flutes and other wind instruments also produce musical sounds – in their case by causing a column of air to vibrate. Drums produce sound when the membrane of the drumhead vibrates. Supposedly Pythagoras' last words to his disciples were, "Work the monochord." Was this his way of saying, "Become a musician," or of saying, "Investigate the nature of the universe?" We understand Pythagoras better if we realize that for him these are the same vocation.

3. Phases of the Moon (448 BCE)

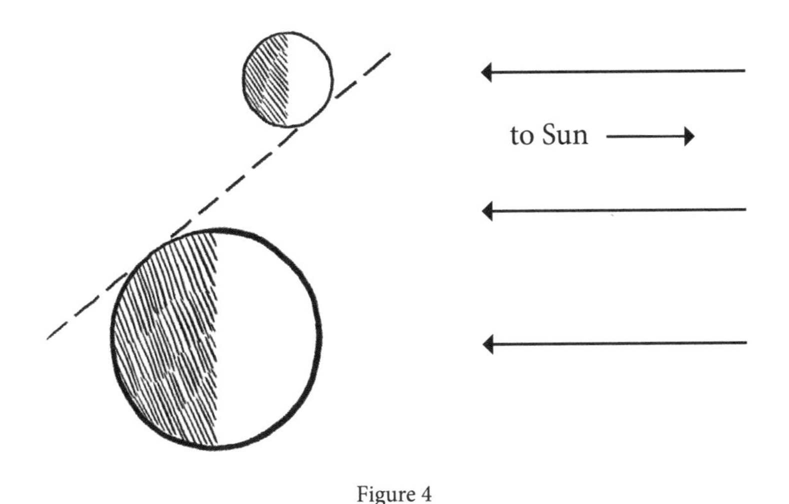

Figure 4

The various appearances of the Moon – new moon (or no moon), tiny crescent moon, quarter moon, gibbous moon (part-way between quarter and full), and full moon – are so familiar we may wonder why they need to be explained at all. Yet a certain kind of mind strives to explain complex phenomena, whether familiar or unfamiliar, in terms of simple concepts. These simple concepts should themselves be plausible and explain other phenomena. If successful, the explanation becomes part of a coherent outlook or theory.

All we need in order to explain the progress of the Moon through an ordered series of phases is to assume that (1) the Moon gives off no light of its own but reflects the light of the Sun, (2) the Moon travels around the Earth in an approximately circular orbit, and (3) the rays of sunlight reaching the Moon and the Earth travel along parallel lines. These ideas are illustrated in figure 4.

But beware. The diagram necessarily distorts other aspects of reality. The Moon is neither so large compared to the Earth nor so close to it. Nor, as is implied, does the Moon pass into the Earth's shadow every month and cause a lunar eclipse or pass between the Sun and the Earth and cause a solar eclipse, for the plane of the Moon's orbit around the Earth is slightly tilted with respect to the plane of the Earth's orbit around the Sun.

The Sun's rays illuminate, at any one time, only half of the Moon's surface and half of the Earth's surface. The rest is in shadow – a shadow that on Earth we call *night*. An observer located in the diagram at the point of contact between the circle of the Earth and the dotted line would just have been carried into this shadow by the Earth's daily anticlockwise rotation. This observer can see only that region of the sky above his local horizon, here indicated with the dotted line. The bulk of the Earth blocks the rest of his view. What this observer sees of the Moon is a relatively thin sliver of reflected light. This sliver appears as a crescent with its horns pointing away from the Sun. As the Earth's rotation continues to carry the observer in an anticlockwise direction, the Moon drops below the observer's horizon.

Figure 5 shows several positions of the Moon, each about seven days apart, as it moves around the Earth. This cyclic motion takes 29.5 days – a *moneth* in Middle English or, as we now say, a *month*. In each position a different portion of the Moon's illuminated surface is visible to observers on the night side of the Earth. These different appearances are the Moon's phases: waxing half moon, full moon, waning half moon, new moon, and all those in between. Since the monthly motion of the Moon around the Earth is slow compared to the daily rotation of the Earth on its axis, an observer on the Earth sees very much the same phase of the Moon throughout any one night.

We cannot be sure who first explained the existence and succession of the Moon's phases in this way. However, we do know that the Greek philosopher Anaxagoras (~500–428 BCE) was the first to leave a written record suggesting important aspects of this explanation. Anaxagoras was a native of Clazomenae, a city of Greek speakers in the middle of the west coast of Asia Minor, now Turkey. Anaxagoras spent 20 to 30 years of his mature life in Athens and so

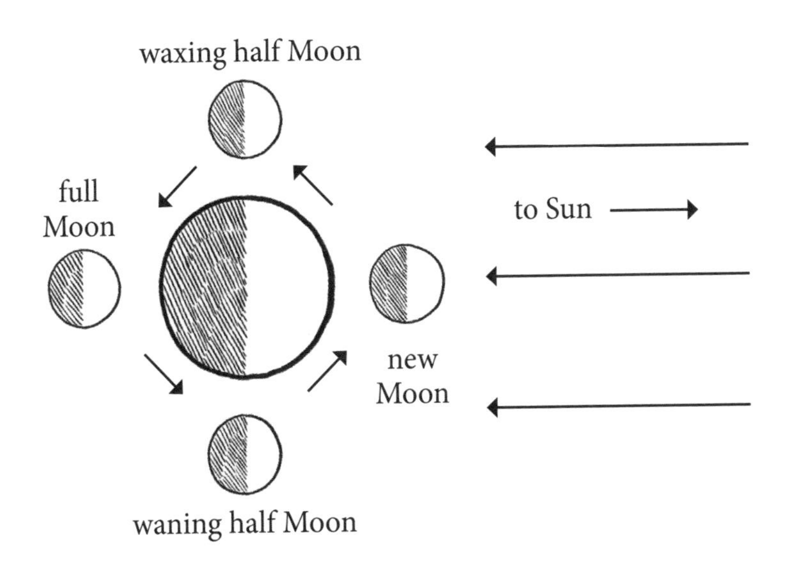

Figure 5

witnessed first-hand the beginning of the Peloponnesian War and other stirring events and impressive achievements of fifth-century BCE Athens. His time in Athens must have overlapped with the lives of near contemporary Athenian tragic playwrights Sophocles and Euripides and with the younger Socrates (469–399 BCE).

We do know that Anaxagoras wrote books because Socrates claimed to have read one of them, though he did not like it very much. Today we have only a few fragments of Anaxagoras's writing preserved as quotations in other ancient texts. "The Sun puts the shine in the Moon" is one of these fragments. Later commentators Aetius and Plutarch, writing four and five centuries after Anaxagoras, claimed that Anaxagoras was the first to clearly explain the cause of the Moon's phases. Diogenes Laertius, who flourished in the third century CE, also claimed that Anaxagoras was the first to put a diagram in a book. If Anaxagoras did, indeed, explain the phases of the Moon with a diagram, I suspect his diagram looked much like figure 5.

Anaxagoras is best known for his creative cosmology: that is, for his way of explaining everything. Anaxagoras's first cosmological principle was that Mind directs and orders all things. Anaxagoras spoke of mind so often that his contemporaries gave him the nickname *Nous*, the Greek word for *mind*, just as today we might, with a little sarcasm, call someone a "brain". It was Anaxagoras's failure to follow through with the idea of Mind that so disappointed Socrates. Instead of explaining phenomena in terms of creation for some purpose as a mindful person might create – that is, for the sake of being beautiful or useful – Anaxagoras, in fact, often resorted exclusively to material and mechanical causes. One such materialistic idea, that the Sun and all the stars are simply fiery pieces of metal, led to his conviction for impiety and banishment from Athens.

4. Empedocles Discovers Air (450 BCE)

Figure 6

The air that surrounds us is invisible, odourless, and tasteless. Nor does it usually produce sound or resist our movement through it. Of course, sometimes we feel a breeze at our back or a wind in our face. Less frequently tornadoes obliterate solid buildings and gale-force winds raise seas that put neighbourhoods under water. No doubt our ancestors had been aware of these phenomena for millennia when Empedocles (490–430 BCE), a native of Acragas in Sicily, sought to explain them. He was one of several physically minded, Greek-speaking philosophers or cosmologists who sought the *principles* or, in Empedocles' usage, the *roots* of all phenomena.

For Thales the single principle was water, presumably because of its ubiquity and because under common conditions water exists in three different phases: solid, liquid, and gas. For the Pythagoreans the single principle was number. For Anaximenes (fl. 500 BCE) it was air. For Heraclitus (fl. 495 BCE) it was fire. In a surviving fragment Heraclitus suggests that while the *logos* (or the account) abides, there is no permanent material thing, since, according to Heraclitus, "You cannot step twice into the same river, for other waters and yet others go ever flowing on."

Empedocles sought to explain both the variableness and the stability of our experience by postulating that everything is composed of just four elements: earth, air, fire, and water. According to Empedocles these four elements are neither created nor destroyed. The mixture and separation of different quantities of earth, air, fire, and water, brought about by the agencies of love and strife, account for the world of change we experience. Much later Empedocles' four elements became the basis of Aristotelian and medieval cosmology.

What prompted Empedocles to consider air as one of the four fundamental elements? Air is a peculiar choice for, unlike earth, fire, and water, air seems bereft of qualities. For instance, we can see the water vapour above a boiling pot, but not the air that supports that vapour. Likewise, we can see the fire under the pot but not the air the fire consumes. One fragment of Empedocles' poem *On Nature*, out of the several hundred of its lines that survive, suggests an answer to this question. In describing human respiration Empedocles compares the pores in our lungs and in our skin with a *clepsydra* – the main part of an ancient water clock that is an open-mouthed jar that can be drained through a small spout at the centre of its bottom. He writes:

> . . . as when a young girl, playing with a clepsydra of shining bronze, puts the passage of the pipe against her pretty hand and dunks it into the delicate body of silvery water, no liquid enters the vessel, but the bulk of air, pressing from inside on the close-

set holes, keeps it out until she uncovers the compressed stream.
But then when the air is leaving the water duly enters.

Figure 6 illustrates the phenomenon Empedocles observed. According to the poem, the young girl inverts the clepsydra, closes its spout or *pipe* with her finger (here replaced with a plug), and submerges its open mouth into "the delicate body of silvery water". Interestingly, something prevents the water from rising inside the clepsydra and assuming the same level as outside. That something is air. For when the child lifts her finger, its seal with the pipe is broken, air rushes through the pipe, and the water below the clepsydra pushes upward. If we have ever doubted the substantiality of air, this little demonstration, which can be reproduced with a kitchen funnel and a sink full of water, should dispel our doubt.

Thales, Anaxagoras, Empedocles, and the other Greek philosophers pre-dating Socrates (469–399 BCE) were imaginative and, for the most part, materialistically minded thinkers who sought to explain phenomena with the fewest, most self-consistent, and most plausible principles. While these cosmologists appealed to common observations (for instance, that water exists in three phases), their tools of discovery and verification were invariably speculation and argument. They did not perform experiments. But one wonders: could Empedocles have resisted imitating the child he observed? Did he not also play with the clepsydra? If so, Empedocles did something unusual for a Greek of his time. He not only thought about nature; he manipulated a natural phenomenon with the intention of learning something new. He performed an experiment.

5. Aristotle's Universe (350 BCE)

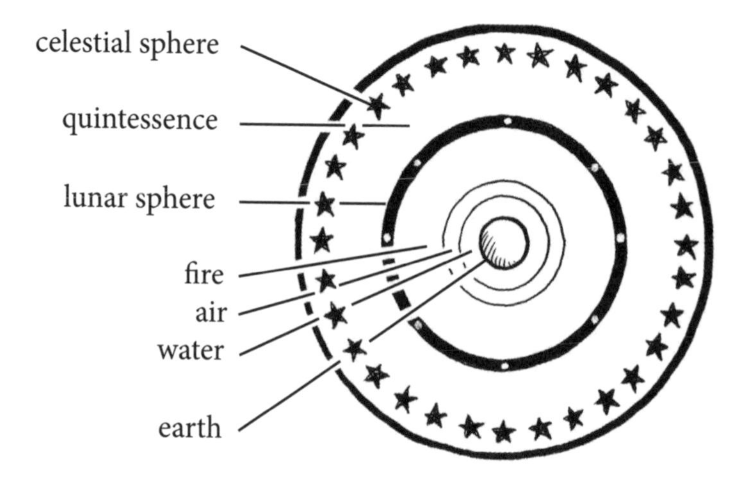

Figure 7

Have you ever heard someone say, "Eventually scientists will figure out how to do it?" You fill in the reference for "it". Travel faster than the speed of light? Build a heat engine with 100% efficiency? Extract energy from the cosmic microwave background? Indeed, it may be that some things once thought impossible will turn out to be quite possible. But it is not true that everything of which we might dream is possible. After all, we live in a world that has a *nature*: a characteristic way of being and of becoming, of remaining the same and of changing.

We may learn about that nature and discover ways to employ it, but we have no power to change the nature of things. According to Francis Bacon (1561–1626), "Nature, to be commanded, must be obeyed." Aristotle (384–322 BCE) transmitted to us this indispensable concept of nature – a concept rejected, perhaps unintentionally, by those who think scientists and engineers can do anything and everything.

The word *nature* comes to us from a Latin root, the Greek equivalent of which is φύσις or *phusis*, from which also derives *physics*. Of course, modern physics was born struggling against certain Aristotelian ideas. Nevertheless, Aristotle's concept of nature is the bedrock upon which the practice of modern physics stands.

Figure 7 illustrates Aristotle's universe – not as one would observe it, but rather in the state of perfection toward which Aristotle's universe tends by virtue of its nature. Earth and water move down toward the centre – earth more persistently than water. Air and fire move up away from the centre – fire more readily than air. Thus upward and downward motions characterize the region below the sphere of the Moon. Objects above the lunar sphere are composed of a fifth substance, the *quintessence* or *ether*. The Sun and wandering stars or planets (not shown in this figure) and the fixed stars reside on transparent spheres that carry them around the Earth in concentric circles. Circular motion characterizes the region above the sphere of the Moon.

Aristotle borrowed many of the features of his universe from his pre-Socratic predecessors: for instance, the four elements (earth, air, fire, and water) and the celestial spheres. Furthermore, the pre-Socratics were the first to formulate the concept of nature. But Aristotle composed these ideas into an ordered whole, a *cosmos*, that answered the questions of his day and, at the same time, remained consistent with commonplace observations.

That last statement needs to be qualified for Aristotle must have observed sublunary objects that do not always move up or down. Toss a clod of earth and it travels along an approximately parabolic arc, at first up, then down, and always in the direction thrown. According to Aristotle, motion requires a mover and if that mover is not the nature of the moving object, motion must be imparted and maintained externally: that is, unnaturally or by "violence".

Thus, it is the hand that throws the clod and the air through which the clod moves that cause its unnatural horizontal movement.

According to this view, to manipulate objects and study their behaviour, that is, to perform experiments, is not a reliable way to study nature. For in doing so one fruitlessly studies that which has no nature: the whimsy of the human boy, for instance, that tossed the clod in a particular way. To manipulate a natural phenomenon is to spoil its naturalness – at least according to Aristotle.

Nevertheless, Aristotle was a great observer of nature and, according to the eminent historian of science George Sarton, "one of the greatest philosophers and scientists of all times". He discovered the law of the lever and was the first to systematically study meteorology. He "carried on immense botanical, zoological, and anatomical investigations [and] clearly recognized the fundamental problems of biology: sex, heredity, nutrition, growth, and adaptation". He structured the elements of logic and originated the inductive method. Aristotle also wrote ageless treatises on literary criticism, ethics, and metaphysics. Indeed, there is hardly a branch of human knowledge to which Aristotle did not contribute.

In 335 BCE Aristotle established a school of philosophy and science in Athens called the *Lyceum*. Those who studied with and followed Aristotle became known as *peripatetics*; that is, those who study while walking from place to place. Aristotle's most famous pupil, Alexander the Great, the son of Phillip II of Macedon, conquered the known world.

According to Aristotle, the realm of the celestial spheres is perfect. Its motions, unlike those of the sublunary realm, are completely natural, manifestly beautiful, and ultimately caused only by the desire for the good. It is not hard to see why Aristotle's view of the universe has influenced thought and literature for over 2,000 years. It is, after all, a privilege and delight to look on the heavens each night and be inspired by their perfection.

6. Relative Distance of the Sun and the Moon (280 BCE)

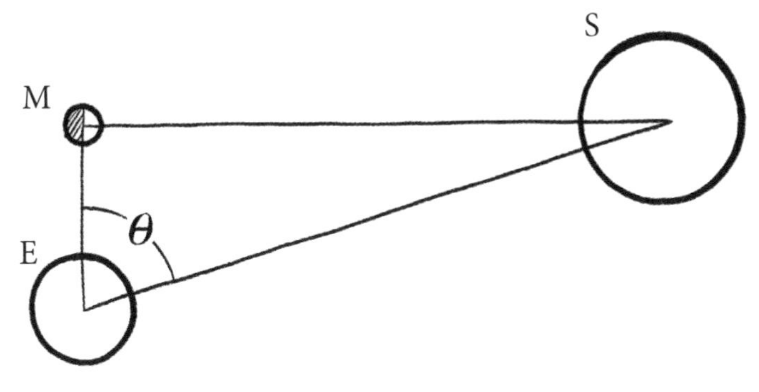

Figure 8

Aristarchus of Samos (310–230 BCE) was the first to determine the relative distance of the Sun and the Moon from the Earth. His method, like that of Thales, depends on the properties of similar or same-shaped triangles. But his application – to the relative distances of heavenly bodies – is much bolder. Aristarchus assumed only that the Moon receives its light from the Sun.

Figure 8 suggests Aristarchus's argument at the cost of making the Sun too close relative to the Earth-Moon separation. When the Moon appears, to an observer on the Earth, exactly half dark and half light – that is, appears as a half Moon – Aristarchus knew that a line from the Sun to the Moon must meet a line from the Earth to the Moon at a right angle. Thus, if at half Moon one is able to measure the size of the slightly-less-than-right angle θ between the Earth-Sun and Earth-Moon lines (admittedly a difficult measurement), one knows all there is to know about the shape of the right-angled triangle formed by Earth, Moon, and Sun. All there is to know about the shape of the right triangle would allow one, for instance, to reproduce a similarly shaped triangle on papyrus and in this way determine the ratio of the two sides representing the Sun-Earth and Moon-Earth distances; that is, the ratio *SE/ME*.

Aristarchus's procedure led to a ratio *SE/ME [=20]* according to which the Sun is about 20 times more distant from the Earth than the Moon is from the Earth. The ratio *SE/ME* is, in fact, closer to 400. Aristarchus's method, however, is sound. At least one scholar has claimed that the techniques available to Aristarchus should have allowed him to be more accurate. If so, perhaps his interest was more with pioneering a new method than with its careful application.

Aristarchus is better known for his teaching that the Earth rotates daily on its axis and revolves yearly around the Sun. But he convinced few of his contemporaries. The problem is that a yearly motion of the Earth around a presumably stationary Sun implies that the relative positions of the stars as observed from the Earth should change during the year. We now know that the closest stars are too distant for this effect of observer motion, the so-called *stellar parallax* effect, to be seen with the naked eye.

Note that figure 8 makes no distinction between Earth- and Sun-centred planetary systems. Both are consistent with Aristarchus's method of determining relative distances since in each case the Moon revolves around the Earth. Thus Aristarchus's contemporaries could, with perfect consistency, accept his determination of relative distances and reject his Sun-centered planetary system.

The manuscript in which Aristarchus made his argument, *On the Sizes and Distances of the Sun and Moon*, refers not only to the *distances* of the Sun and the Moon but also to their *sizes*. His determination of the relative size of the Sun and Moon is also insightful. Aristarchus observed that during a total eclipse of the Sun the disc of the Moon completely obscures the disc of the Sun but just so – as shown in figure 9. Because the larger right-angled triangle appearing in this figure is similar to the smaller one, the ratio of the distances of Sun and Moon *SE*/*ME* from the Earth must equal the ratio of their radii R_S/R_M. Therefore, if the Sun is, as Aristarchus believed, 20 times more distant than the Moon, the Sun must be 20 times larger than the Moon. Since the Sun is, in fact, about 400 times more distant than the Moon, the Sun is actually about 400 times larger than the Moon.

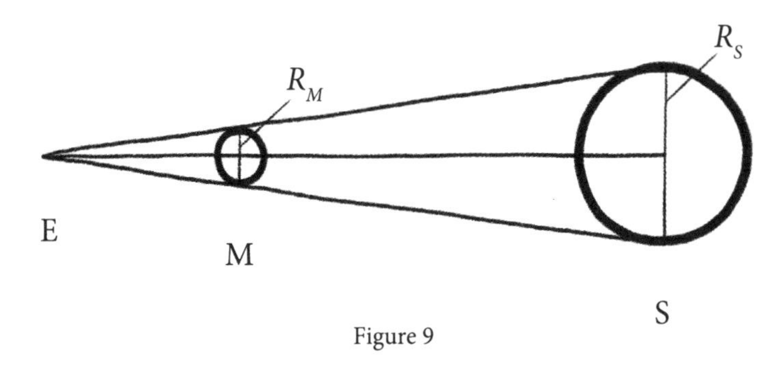

Figure 9

Aristarchus forged yet one more link in his chain of arguments. He noticed that the time required for the Moon to pass into the Earth's shadow during a lunar eclipse is close to the time the Moon stays completely obscured by the Earth's shadow. If so, the Earth's radius must be twice that of the Moon – assuming the Earth's shadow from Earth to Moon is approximately cylindrical. And if the Earth is 2 times larger than the Moon and the Sun is 20 times larger than the Moon, then the Sun must be must be 10 times larger than the Earth. Again, Aristarchus's argument is valid even if his data are not accurate. According to modern measurements the Earth is about 4 times larger than the Moon and the Sun is some 100 times larger than the Earth.

Aristarchus's methods illustrate the intellectual trends of his time. Since he was a younger contemporary of Euclid (the latter flourished around 300 BCE), Aristarchus lived during a period in which the propositions of geometry had became widely known. Subsequently, astronomical knowledge was more frequently framed in geometrical language than before and, in this way, physical science began to distinguish itself from philosophical wisdom. At the same time the centre of Greek learning migrated from Athens to the newly founded city of Alexandria near the mouth of the Nile. While Aristarchus may or may not have travelled from his native Samos to Alexandria, his life falls within the initial stages of a period of cultural ferment set in motion by the conquests and foundations of Alexander the Great.

7. Archimedes' Balance (250 BCE)

Figure 10

Around 300 BCE Euclid organized the mathematical knowledge of his time into *definitions, common notions, postulates,* and *the demonstrations* of *propositions.* Some of the definitions are familiar: for instance, "a line is breadthless length", and some seem a little mysterious: "A straight line is a line that lies evenly with the points on itself." The common notions are self-evident statements common to all kinds of reasoning, such as, "Things which are equal to the same thing are also equal to one other". The postulates are a small group of unproven statements, assumed to be true, such as "All right angles are equal to one other," and the propositions are statements whose truth Euclid demonstrates by valid argument from the postulates, the common notions, and previously demonstrated propositions. The result, contained within the 13 books of Euclid's *Elements,* is an extended deductive system that has, for 2,300 years, been a model of rigorous thinking. The outstanding lesson of Euclid's system is that many truths can be demonstrated and not merely asserted.

Euclid's *Elements* astonishes and charms its readers. It is said that Sir Thomas Hobbes, on first picking up Book I of the *Elements* and reading Proposition 47, the Pythagorean Theorem, exclaimed, "By God, this is impossible." Hobbes then read its demonstration and then the demonstrations of the propositions used to demonstrate Proposition 47 and so on until he had read a good part of Book I *in reverse order* – a method of reading Euclid I do not recommend. On the other hand, I do commend Edna St. Vincent Millay's response to Euclid, a 14-line Shakespearean sonnet, "Euclid alone has looked on Beauty bare", whose middle verses are,

> . . . let geese
> Gabble and hiss, but heroes seek release
> From dusty bondage into luminous air.
> O blinding hour, O holy, terrible day,
> When first the shaft into his vision shone
> Of light anatomized! Euclid alone
> Has looked on Beauty bare. . . .

Archimedes (287–212 BCE), certainly the most original mathematician and physicist of antiquity, also fell under Euclid's spell. We know this because he followed Euclid in organizing what he had discovered about the equilibrium of heavy bodies into a system of postulates, propositions, and demonstrations.

Figure 10 illustrates propositions 6 and 7 of Archimedes' *On the Equilibrium of Planes* that together compose his law of the balance: *two objects balance at distances inversely proportional to their weights* – a law illustrated each time two children of unequal weight balance themselves on a see-saw. The dark line in the diagram is the balance beam, the triangle is the beam's support or pivot, short, light lines mark the beam at equally spaced intervals on either side of the pivot, and the blocks stand for units of weight. In the left-hand diagram a weight of two units is one unit to the right of the pivot and a weight of one unit is two units to the left of the pivot, so that each weight is at a distance from the pivot inversely proportional to its magnitude.

Figure 10 illustrates a demonstration of a particular case of the law of the balance. The demonstration requires only two premises, both quite reasonable. One of these is Archimedes' Postulate 1, *equal weights at equal distances* (from the pivot) *balance*, and the other is a previously demonstrated proposition, Proposition 4, *the centre of gravity of two equal weights taken together is in the middle of a line joining their centres*. The phrase *centre of gravity* refers to the location at which a pair of identical weights can be replaced by a single weight equal in magnitude to the total weight of the pair. Thus, Proposition 4 justifies the transition, shown in the middle diagram above, that makes the stability of the left and right-hand diagrams equivalent. Of course, according to Postulate 1, the weights in the right distribution balance. Therefore, the sequence of diagrams from left to right (or from right to left) demonstrates that a weight of two units located one unit to the right of the pivot balances a weight of one unit located two units to the left of the pivot.

Archimedes' demonstration is more general than ours. Nevertheless, our demonstration exploits the rule justified by his Proposition 4: *a weight at a particular location can be replaced by two weights, each equal to half the original weight, placed equal distances on either side of their original location*. By applying this rule several times one can show that the two balance beams illustrated in figure 11 are physically equivalent. Give it a try. And remember: you are allowed to place blocks on top of the pivot.

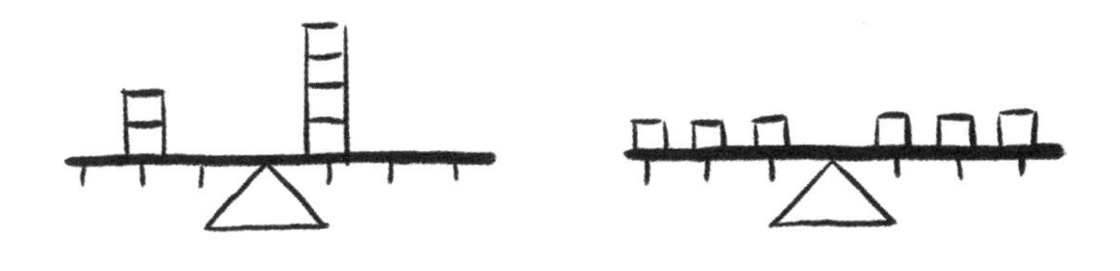

Figure 11

Archimedes may have sojourned for a while in Alexandria, and if so, he may have known the somewhat younger Eratosthenes (276–194 BCE), who plays a role in a later essay. Even so, Archimedes lived the greater part of his life in his native Syracuse, a Greek city on the island of Sicily. Greek colonists had inhabited Sicily and the southeastern coast of the Italian mainland since the 8th century BCE. During Archimedes' lifetime, the Romans extended their dominion over the Italian peninsula and Sicily and engaged the North African city of Carthage in a life and death struggle. Syracuse, and therefore Archimedes, stood directly in the path of this Roman expansion.

8. Archimedes' Principle (250 BCE)

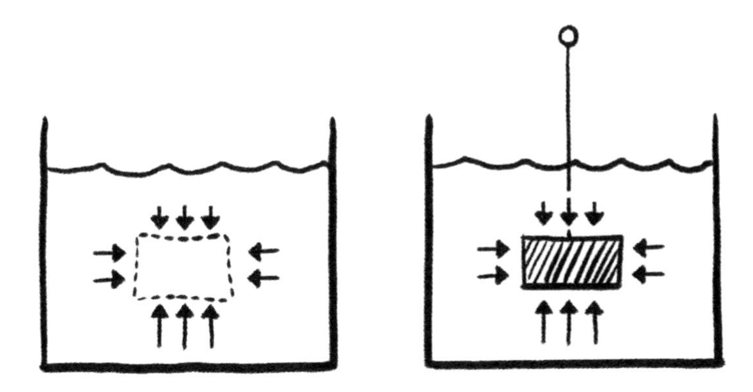

Figure 12

The story goes that when the solution to a particularly challenging problem came to Archimedes in his bath he leapt from the tub shouting *"Eureka! Eureka!"* (I have found it! I have found it!) But what had Archimedes found? According to Vitruvius (ca. 75–15 BCE), the Roman military engineer who told the story almost two centuries after the event, Archimedes had discovered a method for determining whether a crown that had been made for King Hieron of Syracuse was of pure gold, as per instructions, or mixed with silver. The story is a good one – almost too good to be true – for the method Archimedes discovered concerned bodies submerged in a fluid just as Archimedes' body was submerged in his bath water. However, Vitruvius does not tell us the details of Archimedes' method.

Figure 12 illustrates the physics behind what physics teachers call *Archimedes' principle* – the content of which Archimedes outlined in Propositions 3 to 7 of Book I of his text *On the Equilibrium of Floating Bodies*. Archimedes' principle is simply stated and elegantly proved and could have been used to determine the composition of king Hieron's crown.

The left-hand panel of figure 12 shows a container filled with water, or any other fluid, at rest. The dashes outline a region of the fluid, while the arrows represent the direction and magnitude of the pressure exerted by the fluid *outside* the outlined region on the fluid *inside* the outlined region. (The longer the arrows, the larger the pressure.) Note that, as one might expect, the magnitude of this pressure increases with depth. Therefore, the upward push, on the fluid in the outlined region, is larger than the downward push. In fact, the net upward push must be just enough to support the weight of the fluid in the outlined region in order to keep the fluid at rest.

In the right-hand panel of figure 12 the fluid that was in the outlined region has been replaced with an identically shaped object. A string attached to the object keeps it from sinking. Since the fluid outside the object in the right diagram is identical to the fluid outside the region outlined in the left diagram, the net force the outside fluid exerts is also identical. Thus, *the fluid outside an object exerts a net upward force on the object equal in magnitude to the weight of the fluid the object displaces.* This is Archimedes' principle. The argument with which we have reached Archimedes' principle applies to floating as well as to fully submerged objects.

Once we understand Archimedes' principle we can use it to solve problems. Here is one problem physics teachers sometimes assign their students. A cargo ship containing iron ore is in a watertight lock as illustrated in figure 13. The captain orders his crew to dump the iron ore into the bottom of the lock. Does the water level in the lock rise, fall, or stay the same when the ore is dumped? The answer (the water level falls) requires the creative use of Archimedes' principle.

Archimedes also proved that *the surface of a fluid at rest is part of the surface of a sphere whose centre is at the centre of the Earth.* Stop a moment and consider this claim. The surface of every glass of water, every cup of coffee, and every farmer's pond is curved,

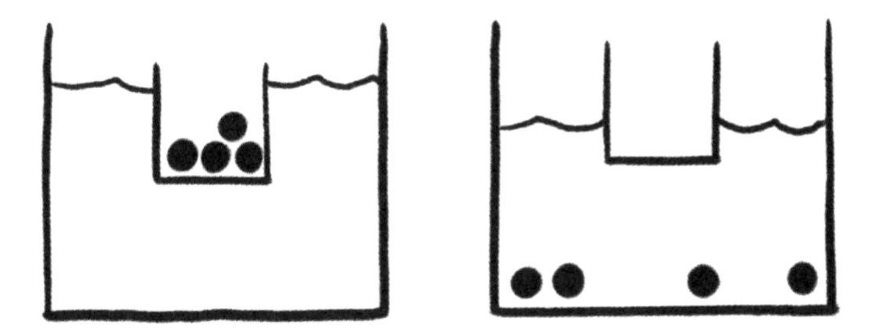

Figure 13

concave downward, with its centre of curvature at the centre of the Earth! Of course, Archimedes' claim ignores the distorting effect of surface tension. But, still, the claim is amazing. The bulk of the Earth pulls on a fluid in such a way as to shape it into a section of a sphere. Figure 14 illustrates this claim and provides the seed of Archimedes's proof, which, however, I do not spell out.

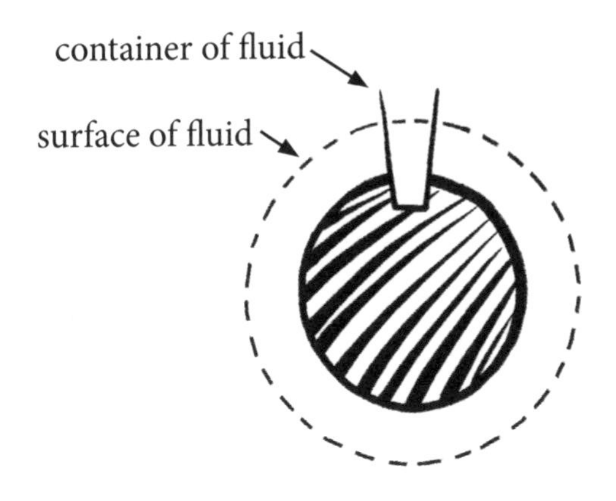

Figure 14

Although Archimedes was primarily a mathematician and physicist, he also invented devices that exploit physical principles: the so-called Archimedean screw that could pump water from a lower to a higher level and the compound pulley that could, in principle, allow one person to, very slowly, lift a massive ship. Some of Archimedes' inventions were weapons of war: for instance, burning mirrors and catapults, which he devised in order to defend his native Syracuse from the Roman army that besieged it in 212 BCE.

But the Romans prevailed and Archimedes died as he had lived – absorbed in a problem of mathematical science. The Roman commander of the besieging army, Marcellus, had given orders that the famous Archimedes, then 75 years old, be spared. But when confronted by an armed Roman soldier Archimedes, who had been studying some figures drawn in the sand, brusquely demanded, "Stand back from my diagrams!" Those were his last words. Evidently, this was no way to address an armed Roman soldier.

9. The Size of the Earth (225 BCE)

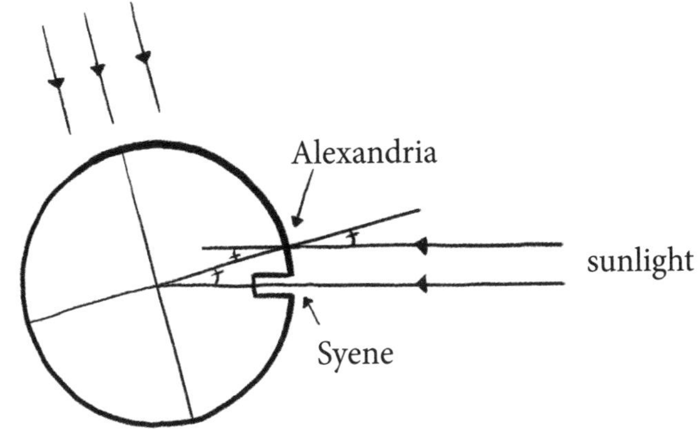

Figure 15

Eratosthenes (276–194 BCE) was born in the North African town of Cyrene (in modern Libya) and educated in Athens, but lived the greater part of his life in Alexandria where around 244 BCE he became the head of its great library. The wealth of this library's holdings can be inferred from the task given Callimachus, a contemporary of Eratosthenes: to catalogue the library's books – and the result of his effort: 120 volumes of bibliography. Thus, we can understand Callimachus's famous complaint, "A great [or large] book is a great evil." Nevertheless, the library and its great books made Eratosthenes' career possible.

Eratosthenes' writings include the text *Geographica*, now lost but often cited in antiquity. This book gathered together what was then known about *geography* – a word he was the first to use in its modern sense. Eratosthenes' compilation of the geographical wisdom of the past into a single, expansive treatise had many imitators during the centuries of Roman domination that followed his death. Pliny's *Natural History*, for instance, included all that was known about the natural world. However, even the best Roman scholars were more concerned with the utility and entertainment value of the learning inherited from the Greeks than with creatively understanding or extending that learning.

But Eratosthenes was an Alexandrian Greek who not only preserved but also built upon the wisdom of the past. For instance, he was the first to add lines of longitude to a map of the known world. On a globe these lines are great circles that pass through both poles. The particular line of longitude or meridian that connects Alexandria and Syene (modern Aswan) plays a role in Eratosthenes' determination of the circumference of the Earth.

Figure 15 illustrates Eratosthenes' method. Eratosthenes noticed that when rays of sunlight reach the bottom of a deep well at Syene, rays of sunlight at Alexandria make an angle equal to 1/50 of a circle with a straight vertical pole or *gnomon*. Eratosthenes also knew that, so distant was the Sun from the Earth, the different rays of sunlight striking the Earth are essentially parallel, and that, according to Euclid, a line falling upon two parallel lines makes the alternate interior angles equal – as illustrated in figure 15. Therefore, according to the geometry of the diagram and Eratosthenes' measurement, an angle equal to 1/50 of a circle (about 7 degrees) with vertex at the centre of the Earth subtends or encompasses the meridian connecting Syene and Alexandria along the surface of the Earth. Consequently, the distance between Syene and Alexandria is 1/50 the circumference of the Earth. It only remained for Eratosthenes to determine this distance and multiply by 50.

As it happens Syene is located near the first waterfall upstream from the Nile's mouth at Alexandria. Between Syene and Alexandria the Nile flooded frequently, and, consequently, was measured frequently by the cadre of Egyptian *geometers*, literally "land measurers," whose job was to preserve the identity of property along the Nile. Eratosthenes had access to the land measurers' records and from them inferred that the distance from Syene to Alexandria was about 5,000 *stadia*. Therefore, according to Eratosthenes' method, the circumference of the Earth is about 250,000 *stadia*.

But how long is a single *stade*? Ancient documents provide at least two different answers to this question. The Egyptian stade is 158 metres, and the more commonly used Greek stade is 185 metres. The first produces a circumference within one per cent of the modern value, kilometres, while the second produces one 17% too large. But comparing Eratosthenes' value to that determined by modern methods teaches us little. It is more important to understand that Eratosthenes' method is sound and based on measurements rather than on speculations.

Eratosthenes also understood that his measurements were uncertain. We know this because Eratosthenes attempted to quantify their uncertainty. For instance, he determined that on the longest day of the year sunlight reaches the bottom of any well at Syene within a circle with radius of about 300 stadia. This effect alone limits the accuracy of Eratosthenes' determination of the circumference of the Earth to plus or minus 6%.

That Eratosthenes assumed the Earth is spherical is unremarkable. For by his time it had long been known that: (1) as one travels north the southern constellations sink toward or below the horizon and the Pole Star rises higher in the night-time sky, and (2) during a lunar eclipse the shadow of the Earth on the Moon is a section of a circle. No observant person could argue with these facts. Only a desire to make sense of them was needed. We know from his determination of the size of the Earth that Eratosthenes had that desire.

Middle Ages

10. Philoponus on Free Fall (550 CE)

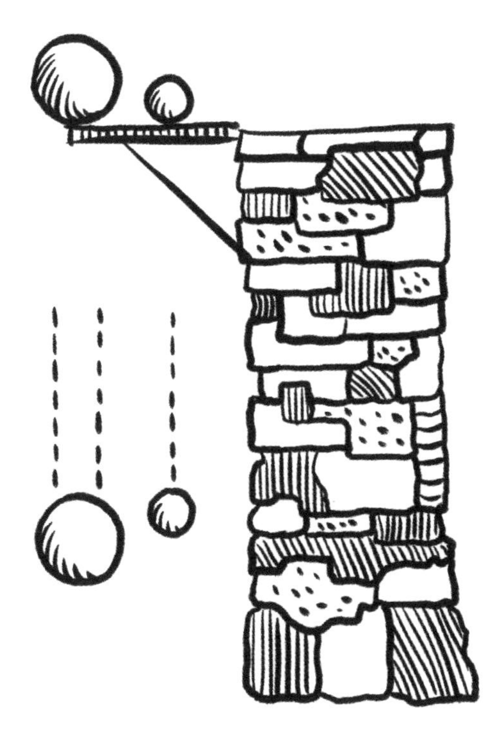

Figure 16

John Philoponus (490–570 CE), whose surname means *lover of toil*, was a Greek Christian who lived and worked as a philosopher, theologian, and scientist in the century immediately following the invasion of Roman Italy by Germanic tribes in 476 CE. While he flourished more than a century after Theodosius had established Catholic Christianity as the official religion of the empire in 380 CE, Philoponus was taught by and worked with pagan philosophers associated with the library in Alexandria. Philoponus wrote extensive commentaries on Aristotle and in several treatises argued against the Aristotelian doctrine of the eternity of the world. He believed that the heavens have the same properties as the Earth and, as well a Christian might, that the heavens are not divine.

Philoponus's analysis of motion critiqued Aristotle's. Aristotle had argued that continuous motion requires either an internal or an external mover in continuous contact with the object moved. Accordingly, the natural downward motion of a heavy object is caused by the object's inner nature and opposed by the air through which it falls. In contrast, the horizontal motion of a projectile is unnatural and requires external movers: at first a mover that initiates the horizontal motion and then the continued push of the air. Philoponus, quite reasonably, doubted that the air could at the same time resist a projectile's natural downward motion and cause its unnatural horizontal motion.

Aristotle also claimed that the time required for an object to fall from rest from a given height is in inverse proportion to its weight. Thus, the heavier an object, the more quickly it should fall. But, according to Philoponus,

> . . . this view of Aristotle's is completely erroneous, and our view may be corroborated by actual observations more effectively than by any sort of verbal argument. For if you let fall from the same height two weights, one many times as heavy as the other, you will see that the ratio of the times required for the motion does not depend [solely] on the ratio of the weights, but that the difference in time is very small. And so if the difference in the weights is not considerable, that is, if one is, let us say, double the other there will be no difference, or else an imperceptible difference, in time . . .

Figure 16 illustrates the situation Philoponus describes. When two objects, one several times heavier than the other, are simultaneously released, the heavier object, according to Philoponus's observation, reaches the ground only slightly ahead of the lighter object – certainly not, as according to Aristotle, several times more quickly.

However, Aristotle's view is not without foundation. Given the difficulty of measuring small time intervals, Aristotle may well have simulated descent in air with descent in water by, for instance, simultaneously dropping heavy and light stones in a pool of clear water. If so, Aristotle would have observed that heavier objects do, indeed, as illustrated in figure 17, fall significantly faster than lighter ones.

We now know that *free fall* – that is, fall through a near vacuum or through a relatively short distance in air – is not comparable to fall through water or oil. In a vacuum all massive objects fall at exactly the same rate just as, for instance, do a hammer and feather dropped together on the surface of our relatively airless Moon. However, in a sufficiently viscous fluid two similarly shaped objects fall at terminal speeds that are, as Aristotle expected, proportional to the object's weight. To observe such descent all one needs is a tall glass of water and two objects of approximately the same shape and size but with very different masses – perhaps a stony pebble and a ball bearing.

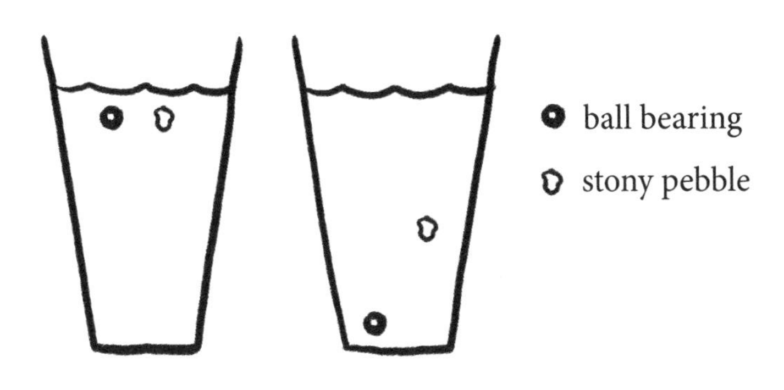

Figure 17

The barbarian invasions that led to the fall of the Western Roman Empire and the subsequent breakdown of Roman institutions disrupted communication between the Latin West and the Greek East. As a result, Philoponus's books and commentaries as well as many other Greek texts became, for centuries, physically and linguistically unavailable to the Latin scholars of the West. Nestorian and Muslim scholars of the 9th and 10th centuries translated many of these texts from Greek into Syriac and Arabic. By 1000 CE this work of translation was largely complete and another translation movement began – this time from the Greek, Syriac, and Arabic into Latin. Witness, for instance, the indefatigable Gerard of Cremona (1114–87) who managed to translate some 70–80 books, including Ptolemy, Aristotle, and Euclid, from Arabic into Latin.

Although Philoponus's books were not translated into Latin until the 14th century, this was well in time for Simon Stevin (1548–1620) and Galileo (1564–1643) to make use of them. Stevin actually reproduced, in Delft in 1586, the free fall experiment Philoponus described. And, while Galileo certainly grasped and exploited the meaning of Philoponus's and Stevin's observations, there is scant evidence that, in similar fashion, he dropped objects from the leaning tower of Pisa. It is an irony of popular history that Galileo is given credit for an experiment he probably did not do and that in his own time was already a thousand years old.

11. The Optics of Vision (1020 CE)

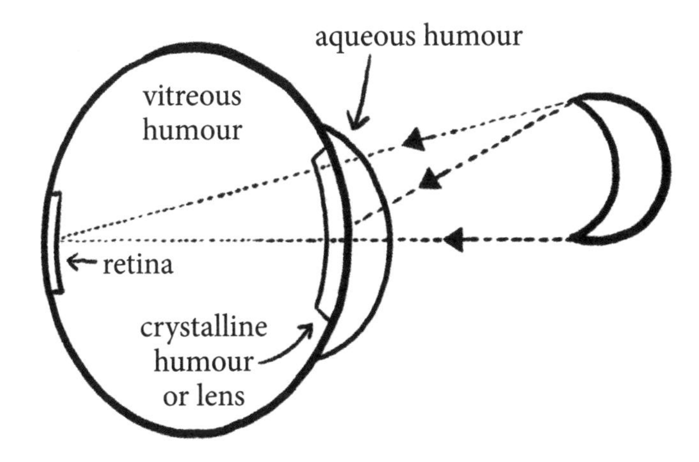

Figure 18

It is said that an Egyptian caliph of the Fatimid dynasty convinced the Muslim sage Ibn al-Haytham (ca. 965–ca. 1040), also known by his Latinized name *Alhazen*, to leave his native Basra in Iraq and come to Egypt in order to design and build a waterway that would regulate the flow of the Nile. Upon close inspection Alhazen found that the project was not feasible. Then, fearing the wrath of the disappointed caliph, Alhazen feigned insanity. This tactic preserved Alhazen's life at the cost of his forced confinement. Even so, Alhazen was able to continue his scholarly work, and, when the caliph died, he recovered his freedom.

During his enforced leisure Alhazen may have occupied himself with copying two works he held in high esteem: Euclid's *Elements* and Ptolemy's *Almagest*. These and other Greek philosophical, mathematical, and medical texts were available to Alhazen in

Arabic translation thanks to the work of 9th and 10th century linguists associated with the *House of Wisdom* in Baghdad.

Greek scholars presented Alhazen with two theories of vision: (1) an *emission* theory, held by Euclid and Ptolemy, according to which rays were emitted from the eye and upon reaching an object rendered it visible; and (2) an *intromission* theory, held by Aristotle and Galen (129 – ca. 200 CE), according to which rays of light travelled in the other direction: that is, from the visible object to the eye. Alhazen knew that very bright objects could damage the eye and so it seemed unlikely to him that, as the emission theory required, the eye could harm itself. He also realized that observing the night-time sky by virtue of rays that travel from one's eyes to the most distant parts of the universe in the instant required to lift one's eyelids was absurd.

Yet the emission theory had one quite useful feature: the rays emerging from the eye (considered as a point) and encompassing the visible object form a cone, the *visual cone*, whose apex is at the eye and whose base outlines the object seen. The visual cone explains, for instance, depth perception. After all, more distant objects of the same size form visual cones with smaller solid angles at the cone's apex. Thus, the angle subtended by a familiar object indicates its distance. How to preserve the useful features of the visual cone and yet adopt the more plausible intromission of rays was Alhazen's problem.

It is important to know that Alhazen laboured under the incorrect assumption that the *crystalline humour* or *lens* was the visually sensitive part of the eye. Not until Kepler dissected an ox's eye in 1604 was it understood that the images responsible for vision are formed on the retina at the back of the eye. Alhazen also supposed, this time correctly, that rays of light emerge in all directions from every point of an illuminated object. For this reason, many rays emerging from a single point enter the surface of the eye at different points and from slightly different directions. For example, consider the two rays coming from the top of the crescent in figure 18. How does the surface of the eye make sense of these two rays and others like them?

Alhazen cleverly, but arbitrarily, answered this question and, in the process, created a theory of vision. He insisted that the surface of the eye is sensitive only to those rays that enter it perpendicular to its curved surface: that is, only to those rays that do not bend or *refract* upon entering the eye. These unrefracted rays form the visual cone. Presumably, refracted rays are, in some fashion, dissipated or rendered incapable of stimulating the lens. In this way the virtues of the visual cone were merged with those of the intromission of rays. In other words, the geometry of Euclid and Ptolemy was merged with the causation and anatomy of Aristotle and Galen.

Alhazen's theory of vision, however flawed we now understand it to be, answered the questions of its day and, as a consequence, was immensely influential. His *Optica* was translated into Latin around 1200. Subsequent contributors to the science of optics, Roger Bacon (1214–1294), Kepler (1571–1630), Snell (1580–1626), and Fermat (1601–1665), all refer to Alhazen.

Alhazen's theory of vision was not his only contribution to optics. He also explained the principles behind the *camera obscura* and understood that when a ray of light leaves one medium, say, air, and enters another, say, water or glass, the incident, reflected, and refracted rays all lie in a single plane – for instance, as represented by the plane contained in figure 19.

Many centuries before Alhazen, Ptolemy (90–168 CE) had made an empirical study of reflection and refraction. Ptolemy correctly surmised that the angle of incidence is always equal to the angle of reflection so that $\theta_{incid} = \theta_{reflec}$. But he also incorrectly proposed that the angle of refraction is directly proportional to the angle of incidence so that $\theta_{refrac} = k \cdot \theta_{incid}$ where the proportionality constant k characterized the different media on either side of the interface. For example, when going from air to water (as in figure 19) Ptolemy found that $\theta_{refrac} = 0.8 \cdot \theta_{incid}$ and so $k = 0.8$. Alhazen showed that Ptolemy's expression describes refraction only through relatively small angles. Not until the 17th century did more generally applicable and accurate theories of refraction become widely available.

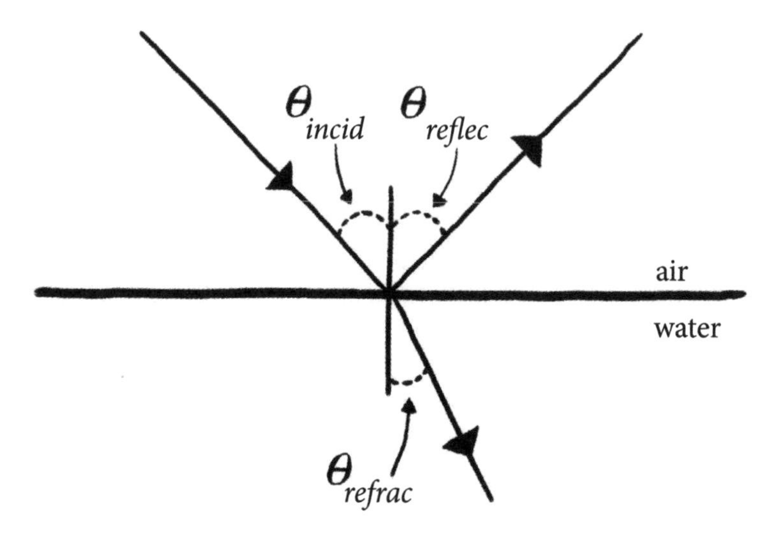

Figure 19

12. Oresme's Triangle (1360)

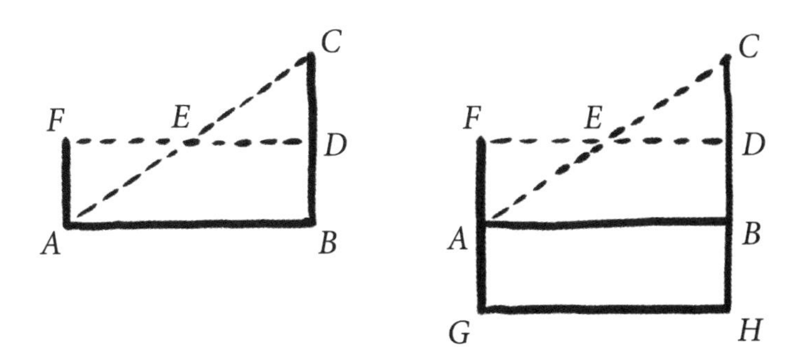

Figure 20

Oresme's triangle, embedded in figure 20, relates two quantities, speed (in the vertical direction) and time (in the horizontal direction), graphically rather than pictorially. It is possibly the earliest such graph. As such it illustrates a proof of a theorem sometimes called the *mean speed theorem* or the *Merton rule* according to which a uniformly accelerated object starting from rest traverses the same distance in a given time as an object moving uniformly at half the accelerated object's final speed. In expressing this proof in graphic language Nicole Oresme (1323–1382), who later became the Bishop of Lisieux in northwestern France, built upon ideas first articulated in antiquity.

The pre-eminent mathematics of Greek and Roman antiquity and of the Middle Ages was geometry, and the preeminent geometry text was Euclid's *Elements*. While the earlier, more familiar books of the *Elements* are non-numerical, Euclid used straight lines to represent numerical magnitudes in the latter books of the *Elements*. The longer the line the greater the magnitude, and a doubly long line has double the magnitude.

While Euclid's lines and magnitudes are abstract quantities without physical reference, Aristotle used straight lines to represent distances before Euclid just as Archimedes and Eratosthenes did, to great effect, after Euclid. After all, that a straight line can represent the distance between two points in space follows naturally from sketching an object extended in space. By dividing a straight line into standard units, the Greeks quantified space just as, by dividing a time interval into so many drops leaving the bowl of a water clock, they quantified time.

The related concept of *speed*, even if composed of distance and time, was not similarly quantified until the period 1325–50 by a group of mathematicians and logicians associated with Merton College at Oxford University: Thomas Bradwardine, William Heytesbury, John of Dumbleton, and Richard Swineshead. These Merton scholars also distinguished among different kinds of motion and investigated their relationships including that identified by the Merton Rule.

Oxford University and its counterparts in Bologna and Paris had grown, in the late 12th century, out of the professionally oriented guilds of masters and scholars, respectively expert in and in need of learning the arts of rhetoric, law, medicine and theology. Contemporaneous with the growth of universities was the recovery and translation into Latin of important Greek and Arabic texts. Thus, the Merton scholars of 1325–50 and a few years later Nicole Oresme at the University of Paris had access to all thirteen books of Euclid's *Elements* and the entire corpus of Aristotle's work.

Oresme's contribution was to translate the largely verbal discussions of the Merton scholars into geometrical language. In the process he also constructed a neat proof of the Merton Rule – a proof that hinges on the distinction between uniform and uniformly accelerated motion. According to a definition, widely circulated in the late Middle Ages, *an object in uniform motion traverses the same distance in equal intervals of time no matter how small the interval.* The Merton scholars constructed a structurally similar definition of uniformly accelerated motion: *an object in uniformly accelerated motion increases its speed by equal amounts in equal intervals of time no matter how small the interval.*

Oresme represented the speed of an object in uniform motion with a series of equal length, vertical lines, separated in the horizontal direction by equal intervals of time as illustrated in the first panel of figure 21. Oresme also represented the speed of an object in uniformly accelerated motion with a series of successively, equally incremented vertical lines separated in the horizontal direction by equal intervals as illustrated in the second panel of figure 21. In both cases, the lines representing speed are perpendicular to a single, horizontal line representing the passage of time. The dashed lines outline the area occupied by the vertical speed lines. The third panel similarly illustrates an arbitrary case of non-uniformly accelerated motion.

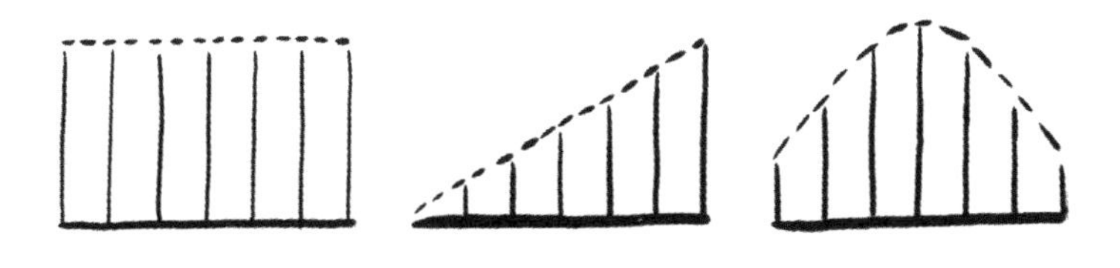

Figure 21

Clearly, the speed lines in the first panel above, representing an object in uniform motion, fill out an area that is proportional to the distance traversed. After all, a car travelling for 2 hours at 90 kilometres per hour has traversed 180 [=2×90] kilometres: that is, the product of the base times the height of the rectangle occupied by its representative speed lines. Oresme assumed, correctly but without justification, that the areas occupied by the speed lines of any kind of motion, non-uniform as well as uniform, also represent the distance traversed – a general theorem that requires the calculus for proof.

Given Oresme's assumption and his use of speed lines, a proof of the Merton rule follows from closely inspecting the triangle ABC in figure 20. Because triangle ABC outlines the speed lines of a uniformly accelerated object, its area represents the distance traversed by that object. By construction, the horizontal line FD bisects the vertical line BC at D so that ABDF is a rectangle with a height BD that represents half the final speed BC of the uniformly accelerated object. Therefore, the area of the rectangle ABDF represents the distance traversed by an object whose uniform speed equals half the final speed of the uniformly accelerated object. In order to prove the Merton Rule Oresme needed only to prove that rectangle ABDF and triangle ABC have the same area.

Today we would simply observe that since the area of any triangle is equal to its base times one-half its height, the area of the triangle in the first diagram of this essay ABC is equal to its base AB times one-half its height BD. But AB times BD is also the area of the rectangle ABDF. Therefore, the rectangle ABDF and the triangle ABC have the same area.

However, Oresme's proof was closely based on Euclidean propositions. Accordingly, note that the vertical angles CED and AEF are equal (Proposition 15 of Book I of Euclid's *Elements*) as are the right angles AFE and CDE (Postulate 4). Therefore, the remaining angles FAE and DCE of the two triangles, CDE and AFE, are also equal. Furthermore, the sides CD and FA are equal because, by construction, the horizontal line FD bisects the vertical line BC at D. Therefore, the triangles CDE and AFE are equal in area (Proposition 26). Adding these equal-area triangles to the same quadrilateral ABDE produces two differently-shaped but equal area figures: the triangle ABC and the rectangle ABDF (Common Notion 2). Since these figures represent the distances traversed, respectively, by a uniformly accelerated object and by an object with uniform speed equal to half the final speed of the accelerated object, the Merton rule is proved. The diagram in the right panel of figure 20 merely extends the proof to allow for objects with non-zero initial speed.

While this analysis may appear wordy and inefficient, even obscure, to those habituated to the algebraic methods of today, our aim here is to understand the medieval science of motion rather than to judge it. And to understand Oresme's analysis we need to reproduce Oresme's pattern of thought.

The Merton scholars' distinction between uniform and uniformly accelerated motion, their discovery of the Merton rule, and Oresme's geometrical proof of it, are exercises in *kinematics*: that is, exercises in the description of motion rather than an exploration of the *dynamics* or the causes of motion. While the dynamics of the 14th century remained enmeshed in Aristotelian concepts, the kinematics of the Merton scholars and of Oresme was a real advance on Aristotle. In due course, Galileo (1564–1642) restated Oresme's proof of the mean speed theorem under the heading "Naturally Accelerated Motion" in the "Third Day" of his text *Two New Sciences*, and René Descartes' (1596–1650) invention of what we now call *Cartesian coordinates* made explicit what Oresme's graphical analysis merely suggested.

13. Leonardo and Earthshine (1510)

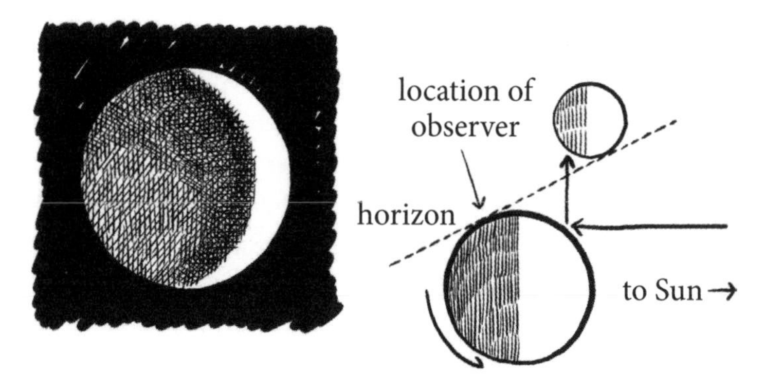

Figure 22

If this text had a section called "Renaissance Science", Leonardo da Vinci (1452–1519) would be its exemplar. Yet, while Leonardo was of the Renaissance, he was not the kind of scholar glorified by the humanists of his time: a scholar nurtured in classical history and literature, having perfect Latin, skilled at rhetoric, and able to speak with confidence at public gatherings. Rather, Leonardo's education was incomplete, his Latin was poor, and he had little interest in public affairs. But Leonardo was a keen observer of nature, an avid experimentalist, and drawn to practical applications. While the classically educated scholars of the Italian Renaissance quoted *authors*, Leonardo cited *experience.*

Leonardo poured much of his experience into 13,000 notebook pages of drawings and text – pages that have enriched the language of visualization. He originated the aerial view so helpful in topography and mapmaking and the idea of presenting different sides of the same object – for instance, of the aorta of an ox. He pioneered the use of anatomical cross-sections, and observed that, at the same distance, a bright object appears larger than a less bright object of the same size.

Leonardo may have intended his notebooks to comprise a profusely illustrated encyclopedia of technical knowledge. But, as they come to us, their pages have no order other than that imposed by the vicissitudes of Leonardo's life. He usually wrote from right to left with characters slanting leftward: so-called mirror-image cursive. We do not know whether this practice was meant to preserve the privacy of his entries or was simply more convenient for the left-handed Leonardo.

The notebooks do, however, help us understand how Leonardo could be a prolifically inventive genius and yet have so little influence on the development of science. Like Archimedes he focused on isolated problems. But, unlike Archimedes, Leonardo failed to develop collections of coherent ideas that explain more than the subject at hand. It is as if the very fertility of his mind and the concreteness of his artistic vision fragmented his scientific efforts and, in this way, kept him from developing powerful and abstract theoretical explanations. Even so the fragments of his thought are often intriguing. Figure 22 illustrates one of them, on *earthshine*.

When the Moon is a waxing or waning crescent, the shaded, relatively dark surface between the horns of its crescent glows with a faint, ghostly light – as suggested in the left panel of figure 22. Leonardo's explanation of earthshine, illustrated in the right panel, is the earliest documented explanation of this phenomenon. According to Leonardo, a significant part of the sunlight striking the Earth is reflected from its surface. The fraction of sunlight that reflects from the Earth's surface, known as its *albedo*, is close to 30%. Some of this reflected light strikes the dark side of the Moon, and some of that light is reflected back to the Earth and observed as earthshine.

Leonardo got one detail of his explanation wrong. He believed that sunlight reflects primarily from the Earth's oceans, in particular from the tops of ocean waves. In fact, the Earth's clouds reflect much more sunlight than do its oceans. Photos taken from orbiting spacecraft confirm that the brightest parts of the Earth are its cloud-covered areas. And when the Earth's cloud cover changes, the albedo of the Earth also changes. In contrast, the Moon has virtually no atmosphere and its albedo, about 12%, remains constant in time. Therefore, measuring changes in the intensity of the earthshine is equivalent to measuring changes in the Earth's albedo. The latter has become an important input to climate change models.

When walking the streets of Florence and Milan Leonardo carried a notebook in which he sketched whatever caught his attention: people, buildings, and landscapes. On occasion he would follow a stranger for hours until he could rough out their visage on paper. Leonardo also drew what he could only imagine: flying machines, cannons that shot exploding shells, and shoes that allowed one to walk on water. He designed a car powered by two sets of springs. While one set unwound and propelled the car forward, the car's passenger would wind the other set. He envisioned a rotating spit powered by the same fire that cooked the flesh impaled upon it and a house of prostitution with an unusually large number of doors. Many of his designs are of practical devices that would, in time, be built. But no one has yet constructed Leonardo's wake-up device contrived of mechanical relays that, when triggered by a water clock, jerked a sleeper's feet into the air.

Leonardo was also keenly interested in mathematics and prepared illustrations for a mathematical text, *De Divina Proportione* (1509), written by his friend Luca Pacioli. But, of course, Leonardo is most famous for his paintings – above all for *The Last Supper* and *The Mona Lisa* – paintings with animated postures; expressive faces; modelled, pointing hands; and pyramidal compositions. Leonardo painted in oils with a broader range of light and dark shades than is usually seen with the eye – a technique art historians call *chiaroscuro*. It may have been that the artist in Leonardo was drawn to the chiaroscuro of earthshine while the scientist in him sought its explanation.

Early Modern Period

14. The Copernican Cosmos (1543)

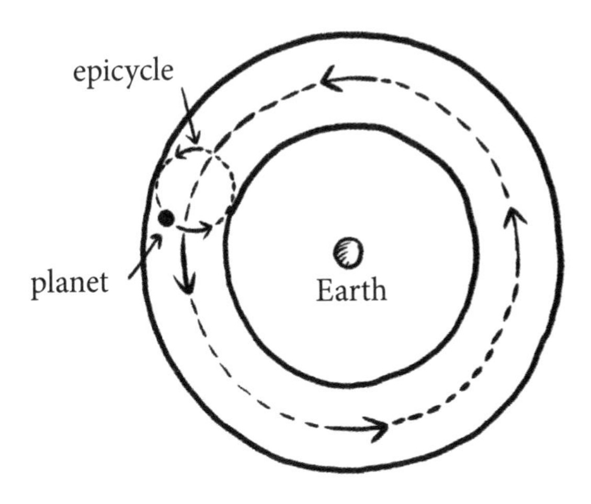

Figure 23

The task of ordering the visible universe into an intelligible whole or *cosmos* has long challenged astronomers. According to Aristotle (384–322 BCE), the heavenly bodies – Moon, Sun, wandering stars or planets, and fixed stars – are embedded in rigidly rotating, Earth-centred, transparent spheres. In this way each "star" moves uniformly in a circle around a stationary Earth. Ptolemy (90–168 CE), who was a great observer of the heavens, embellished Aristotle's basic structure in order to better account for what he actually saw: planets whose brightness varied and planets that sped up, slowed down, and sometimes reversed direction.

Figure 23 illustrates how Ptolemy's embellishments apply to a single planet. Accordingly, the planet moves in a relatively small circle or *epicycle* whose centre, in turn, moves along the primary circular orbit or *deferent*. Thus, two concentric spheres border the epicycle's orbit and determine its size. The epicycle itself accounts for the planet's occasional backward or *retrograde* motion. The centre of the epicycle's bordering spheres can, in turn, be shifted from the centre of the universe where the Earth resides – a feature not illustrated here. Ptolemy's task was to derive numbers that characterize these circles, motions, and shifts from known sequences of observed positions in order to produce an empirically accurate geocentric model of the cosmos. Such was his success that astronomers, astrologers, and calendar makers found his model useful for more than 1,400 years.

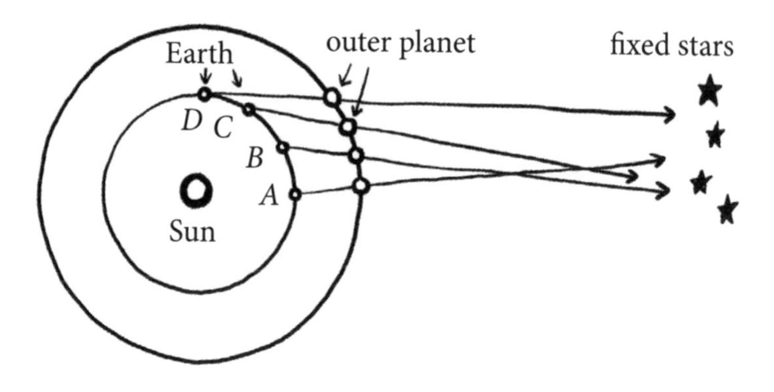

Figure 24

Other ancient astronomers, notably Aristarchus (310–230 BCE), placed the Sun at the centre of the universe, but because such *solar systems* implausibly require a moving Earth they never gained many adherents. Not until 1543 when Copernicus (1473–1543) published his *On the Revolutions of the Celestial Spheres* did the *heliocentrism* of Aristarchus begin to replace the *geocentrism* of Aristotle and Ptolemy. Yet one wonders. What advantage did the cosmos of Aristarchus and Copernicus have over the cosmos of Aristotle and Ptolemy? After all, because Copernicus allowed only spheres that were centred on the Earth, he had to employ even more epicycles than did Ptolemy in order to achieve an equivalent accuracy. The answer is that Copernicus's cosmos *requires* retrograde motion, while Ptolemy's merely *allows* for it. Copernicus's system had a logical coherence that Ptolemy's lacked.

Figure 24 illustrates this logical coherence. Four straight lines of sight from the Earth to an outer planet and on to the "fixed" stars in its background are shown. Because Copernicus arranged the planets so that the more quickly moving ones are closer to the Sun, the connected pairs of small open circles representing contemporaneous positions are further apart on the Earth's orbit than they are on the outer planet's orbit. Therefore, as the Earth moves from point A, near opposition, to point B the outer planet appears to move backwards or to retrogress relative to the background of fixed stars. At C this retrogression diminishes and at D the outer planet's forward motion resumes.

Retrograde motion is observed every time an outer planet nears *opposition*: that is, every time Sun, Earth, and outer planet line up in that order. Of course, Ptolemy's geocentric cosmos also accounts for retrograde motion but only with just the right, individually chosen epicycles, non-concentric spheres, and planetary speeds.

Copernicus must have had the connection between the real motion of the Earth and the apparent retrograde motion of the outer planets in mind in writing his introduction to *On the Revolutions.*

> I finally discovered . . . that if the movements of the other wandering stars are correlated with the circular movement of the Earth, and if the movements are computed in accordance with the revolution of each planet, not only do all their phenomena follow from that but also this correlation binds together so closely the order and magnitudes of all the planets and of their spheres or orbital circles and the heavens themselves that nothing can be shifted around in any part of them without disrupting the remaining parts and the universe as a whole.

After studying in Krakow, Bologna, Rome and Padua, Copernicus performed the duties of a trustee, physician, translator and diplomat for the diocese of Varmia in so-called Royal Prussia. Copernicus's native language was probably German even though this region was then a part of the kingdom of Poland and today, after many vicissitudes, is again part of Poland, now a republic. He saw the birth of the Protestant movement launched by Martin Luther in 1517 and the ravaging of his home by Teutonic Knights. Against the wishes of his bishop Copernicus extended hospitality to the Lutheran mathematician Joachim Rheticus. Rheticus successfully urged Copernicus to publish *On the Revolutions*.

Copernicus held the first printed edition of *On the Revolutions* in his hands shortly before his death in 1543. We do not know whether Copernicus was aware of the anonymous preface, inserted by another Lutheran mathematician, Andreas Osiander. If so, he would have been dismayed. For Osiander's preface characterized heliocentricism as a mere calculational device that allows one to make accurate predictions without pretending to describe reality. Yet it is clear from the introduction to *On the Revolutions* that Copernicus believed in the reality of the heliocentric universe. Copernicus believed that he had discovered the "machinery of the world", which had been constructed by the "Most Orderly Workman of all".

15. The Impossibility of Perpetual Motion (1586)

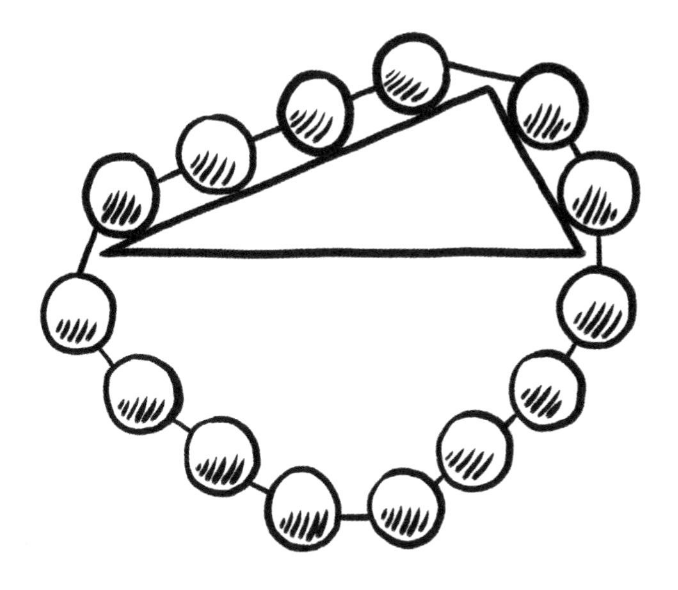

Figure 25

A chain of fourteen identical, spherical beads is draped over a triangular support whose lower edge is parallel to the ground. According to Simon Stevin (1548–1620), its Flemish originator and contemporary of Shakespeare (1564–1616), this *clootcrans* or "wreath of spheres", as it has been variously called, must remain stationary even if one assumes the beads can slip without friction on their supporting surface. Suppose, Stevin reasoned, that the wreath were to slip clockwise. Each sphere would soon take up a position previously held by an adjacent sphere. Then the wreath would recover its original aspect, and then slip again, and so on, *ad infinitum*. Since perpetual motion is clearly absurd, the wreath of spheres must remain in its original position.

The Nobel laureate Richard Feynman referred to Stevin's wreath, in his *Lectures on Physics*, in order to highlight the impossibility of perpetual motion. Feynman went on to say of Stevin's wreath that, "If you can get an epitaph like that on your gravestone, you are doing fine." The impossibility of perpetual motion is, indeed, an important physical concept. Sadi Carnot appealed to it in 1824 in order to motivate his statement of the second law of thermodynamics. But we know neither where Simon Stevin is buried nor the location of his gravestone. Feynman must have meant, not Stevin's gravestone, but rather the memorial statue of Stevin designed by Eugène Simonis and erected in Bruges in 1846 at a place now called *Simon Stevin Plaza*. This statue shows Stevin holding a scroll upon which the wreath of spheres is engraved. Simonis, no doubt, took the wreath of spheres from the title page of Stevin's text *The Principles of the Art of Weighing.*

In this text Stevin argues that since that part of the chain hanging below the supporting triangle is symmetrically arranged around a vertical line passing through its centre, one can remove that part of the chain without disturbing the equilibrium of the wreath's remaining parts – as shown in the left diagram of figure 26. One can also replace the several beads on each inclined plane with a single bead of the same total weight without disturbing that equilibrium – as illustrated in the right diagram of figure 26. (I have, without modifying anything essential, added a frictionless pulley to the arrangement.) These transformations prove a theorem according to which: *Two weights on inclined planes balance each other if the magnitude of each weight is proportional to the length of the inclined plane upon which it rests.* This theorem had been discovered much earlier by one Jordanus Nemorarius, a figure about whom little is known – only that he wrote in Latin and flourished somewhere between 1050 and 1350. But it is Stevin's method of proof, starting as it does from the impossibility of perpetual motion, that interests us.

Figure 26

As a young engineer Stevin helped design and build wind-driven mills that drained the swamps of his native Flanders, now northern Belgium and the adjacent parts of Holland. Eventually Stevin was drawn into the struggle against the Spanish domination of the Low Countries – a struggle in which Stevin served Prince Maurice, the son of William of Orange, as tutor and military advisor. As Maurice's military advisor Stevin brought rational principles to the laying of sieges, the building of fortifications, and the supplying of armies. As Maurice's tutor Stevin compiled textbooks on dialectic, arithmetic, geometry, algebra, mechanics, astronomy, and music – textbooks that contain not only what was then known but also Stevin's contributions to each subject. So much did Prince Maurice value Stevin's textbooks that he carried them with him on campaign.

Stevin is also famous for enthusiastically, if not always convincingly, promoting the use of his native Dutch. According to Stevin, the Dutch language is particularly suited to the scientific enterprise. For Stevin supposed that an efficiently constructed language should represent each single thing by a single word composed of a single syllable. And, according to his investigations, Dutch has more monosyllabic words available for this purpose than either Greek or Latin or, presumably, any of their derivative languages. Stevin even imagined Dutch to be the language of a mythical ancient and enlightened Golden Age in which all people lived in peace and prosperity. Such were his fancies, but Stevin did materially contribute to the evolution of the Dutch language by coining new Dutch words for newly developed technical concepts.

Stevin's efforts to promote Dutch were a part of a larger movement toward employing the vernacular in scientific writing – a movement that attracted a new class of readers to the literature of science.

16. Snell's Law (1621)

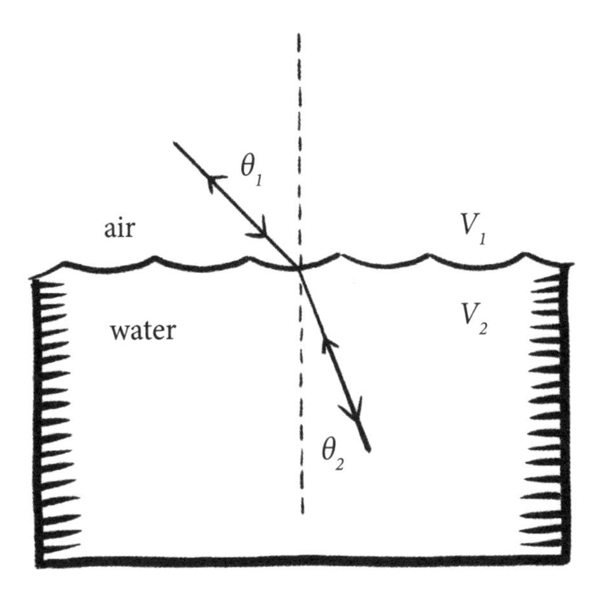

Figure 27

One of the most familiar manifestations of *refraction* is the broken appearance of a straight object resting in and projecting out of a glass of water. While the geometry of this particular phenomenon (involving as it does the light reflected from the object seen, its propagation through water and air, and its reception at the eye) is quite complicated, the essence of refraction is simple.

In a homogeneous medium composed of, for instance, water or air, light travels along straight lines. But when a beam of light or *light ray* leaves one medium and enters another, that part of the light not reflected at the boundary *refracts* or bends toward or away from the line *normal* (or *perpendicular* in a two-dimensional view) to the boundary between the two media. In particular, a ray inclines toward the normal when leaving air and entering water and inclines away from the normal when leaving water and entering air – as illustrated in figure 27 and in figure 19 of essay 11.

Ptolemy (90–168 CE), Alhazen (965–1040), and Kepler (1571–1630) all sought and failed to find an accurate mathematical description of refraction. Not until around 1621 did such a description, attributed in correspondence of the time to the Dutchman Willebrord Snell (1580–1626), emerge. Snell's law asserts that the angles, θ_1 and θ_2, between the ray and the normal to the boundary in each medium, are related to each other by the equation $\sin\theta_1 / \sin\theta_2 = n_2/n_1$ where the indices of refraction n_1 and n_2 characterize the two media. Snell and his contemporaries could, by measuring the two angles θ_1 and θ_2, determine the ratio of one index to the other. When medium 2 is water and medium 1 is air the ratio n_2/n_1 is about 4/3.

Just as important to us as the accuracy of Snell's law is what its form says about the nature of light. René Descartes (1596–1650) demonstrated that Snell's law follows from the hypothesis that light is composed of tiny particles that upon crossing the boundary between two different media either speed up or slow down in the direction of the boundary normal. When, for instance, leaving air and entering water, the particles of light speed up – at least according to Descartes. Given this hypothesis and that $n_2/n_1 = 4/3$, light is faster in water than in air by a factor of 4/3.

Descartes' interpretation of Snell's law is clever, but is neither compelling nor unique. Pierre de Fermat (1601–1665) reasonably objected that, on the contrary, light should travel more slowly in water than in air, since water, being denser than air, must offer more resistance to the particles of light. Elevating this supposition to a postulate, Fermat found that Snell's law follows from an ingenious principle of his own invention: light travels between two points along the quickest route – a principle now known as *Fermat's principle* or the *principle of least time.*

Compare, for instance, the path taken by light as it travels (from air to water) across the air-water interface, as shown in figure 27, to the path taken by a lifeguard as she runs along the beach, plunges into the water, and swims to a person in distress. Since the lifeguard can run faster on the beach than she can swim in water, she minimizes her travel time by covering more distance on the beach than in the water. Consequently, when entering the water she bends her direction of travel toward the person in distress and, thus, toward the perpendicular to the beach-water interface.

Of course one could decide between Descartes' and Fermat's interpretations of the phenomenon by measuring the speed of light in water and comparing it to the speed of light in air. If light is faster in water than in air, Descartes is right; if slower in water than in air, Fermat is right. But technical difficulties delayed such measurements until 1850, at which time it was shown that light travels more slowly in water than in air. Fermat was right.

In the meantime the followers of Descartes sharply criticized Fermat's principle as unphysical. They asked, "Is light supposed to try out all possible paths, compare their transit times, and then choose the quickest path?" It is not so surprising that lifeguards can discern the quickest path and also that ants can and do search out

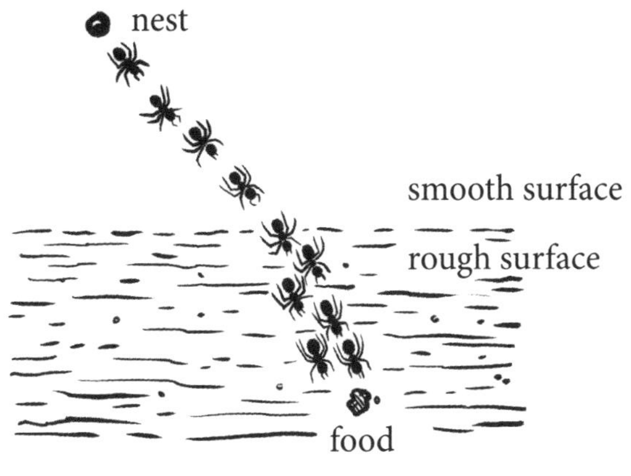

Figure 28

and occupy the quickest route between their nest and a supply of food, as illustrated in figure 28, for in both cases theirs is a learned, goal-oriented behaviour. To the followers of Descartes and most other 17th century natural philosophers, as the scientists were then called, light propagation had to be a purely mechanical process whose explanation necessarily excludes such behaviour. It was not until the wave theory of light triumphed in the early 19th century that the conflict between a mathematically sufficient description (based on Fermat's principle of least time) and the expectations of a mechanistic physics was resolved.

17. The Mountains on the Moon (1610)

Figure 29a

By the first decade of the seventeenth century the time had come for the telescope to be invented. In Holland several Dutch lens grinders and spectacle makers hit upon the same idea at the same time: a tube that aligned two lenses, one concave eyepiece and one convex light-gathering objective. In 1608 one of these Dutchmen, Hans Lippershey, attempted to patent a spyglass, as it was then called, that enabled one to see "things far away as if they were nearby". Lippershey's spyglass only magnified linear dimensions by a factor of three. Even so, since it had obvious military applications, the news of its discovery spread quickly across Europe.

Galileo Galilei (1564–1642) heard of the spyglass in May of 1609, discovered the principle of its construction, and that summer began making his own improved spyglasses. Eventually he achieved a magnifying power of 30. This number is important for, as Galileo later reported, one needs a magnifying power of at least 20 to see the astonishing sights he saw when he first pointed his spyglass toward the heavens: the Moon's surface not smooth, as had been supposed, but rough with mountains and craters; the Milky Way resolved into numerous individual stars; and, most amazing of all, four new wandering "stars" orbiting Jupiter.

Galileo understood the importance of these discoveries. Not only were they novelties, intrinsically interesting, and easily comprehended, they also had far-reaching consequences for our understanding of the cosmos. And because Galileo wanted to quickly communicate these discoveries to the scholars of Europe, he uncharacteristically wrote his short report, the *Sidereus Nuncius* or *Starry Messenger*, in the Latin of his day rather than in the vernacular Italian. Even so the *Starry Messenger* created an immediate sensation in Italy and was discussed on the streets of Padua, Venice, Florence and Rome.

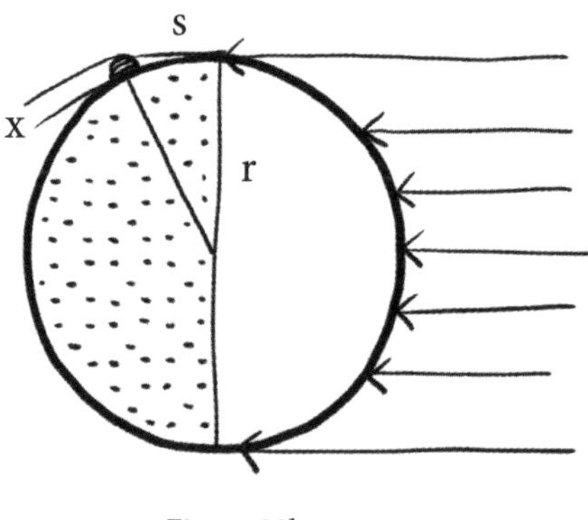

Figure 29b

Figure 29a is by Galileo's own hand. It shows the surface of the Moon as he saw it, roughened with mountains and craters. Figure 29b illustrates the method he used to determine the relative height of these mountains. Its key is that Galileo recognized in the Moon phenomena similar to phenomena on Earth. In particular, he recognized that the white spots just to the left of the line dividing the Moon's shaded and illumined parts in figure 29a are mountaintops ablaze in the rising or setting Sun. Figure 29b shows how the Pythagorean theorem $(r+x)^2=r^2+s^2$ relates the radius of the

Moon r, the height of a mountain x, and the distance from an illuminated mountaintop to the line dividing the shaded and lighted lunar hemispheres s. Because Galileo had a value for the Moon's radius r and could estimate the ratio r/s from his drawing, he was able to use this equation to find values for the height of various lunar mountains. He found the highest of these mountains to be about 4 miles high – not far from modern determinations.

The very roughness of the surface of the Moon undermines the distinction, so important in Aristotelian cosmology, between the imperfect sub-lunar realm of the Earth and its atmosphere and the perfect realm of the heavens, the Moon included. At the time of his telescopic discoveries Galileo was a secret Copernican. Within a few years he would become a public advocate of the Copernican cosmos. It seems likely that these discoveries, especially of the rough surface of the Moon and of the four orbiting satellites of Jupiter, confirmed and energized an advocacy that, eventually, led to Galileo's conflict with the Church.

Galileo's *Starry Messenger* is itself a jewel of composition. Historians of science value it highly. Scientists and science writers should also. Galileo's later writing, found for instance in the dialogue *Two Chief World Systems*, is famous for its sarcastic polemics that destroyed his opponents' positions. But in the *Starry Messenger* we find no sarcasm and no polemics. Rather we find apt metaphors, rounded sentences, and efficient summaries that clearly express complex ideas with excitement. The *Starry Messenger* is a delight for readers and a model for writers.

18. The Moons of Jupiter (1610)

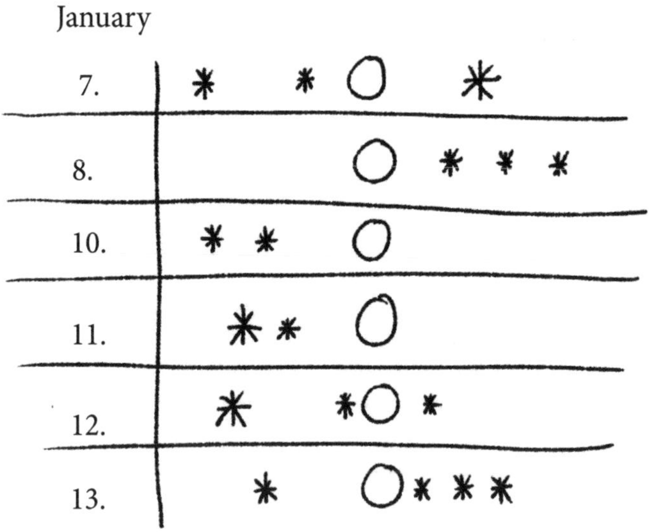

January

7.

8.

10.

11.

12.

13.

Figure 30

Among the marvels Galileo saw when he turned his newly constructed telescope toward the heavens in the winter of 1609–1610 were the four brightest moons of Jupiter. He dubbed these the *Medicean planets* in order to flatter the man whose patronage he sought and to whom he dedicated the 70-page booklet he published in March of 1610 that reported on these telescopic discoveries. This was the *Sidereus Nuncius,* that is, the *Starry Messenger*, dedicated to "The Most Serene Cosimo II de' Medici, Fourth Grand Duke of Tuscany".

In this booklet, Galileo described the way in which he constructed a telescope capable of magnifying linear dimensions by a factor of 30 and the things he saw with his telescope: mountains on the

Moon, newly visible stars into which he resolved the Milky Way, the phases of Venus, and the finite-sized, disc-like appearance of the planets. Then he announced that,

> There remains the matter which in my opinion deserves to be considered the most important of all – the disclosure of four PLANETS never seen from the creation of the world up to our own time, . . .

On the night of 7 January 1610 Galileo observed that Jupiter appeared close to what he initially took to be three small stars in its background. But he also noticed, and later recalled, that these stars and Jupiter unaccountably lined up along the ecliptic: that is, along the band in which the planets move through the background of "fixed" stars. That night two of these stars were to the east of Jupiter and one to the west – as illustrated in figure 30. The next night, 8 January, all three were to the west of Jupiter and again lined up along the ecliptic. Although these appearances interested Galileo, he did not yet understand that these "stars" were Jupiter's moons.

Because the sky was overcast on 9 January, Galileo made no observations. Then on 10 and 11 January only two stars appeared, on both occasions to the east of Jupiter. When Galileo began observing on 12 January he again saw only two stars: the brighter one to the east of Jupiter and the less bright one to the west. But as he was observing another star emerged from the east side of Jupiter. On 13 January Galileo saw four stars – all lined up with Jupiter along the ecliptic.

Galileo kept observing every clear night for two months, long enough to conclude that there were four such stars, moons, or planets, as he variously called them, illuminated by the Sun, and that they revolved, in unequal circles, around Jupiter, but not long enough to determine their periods of revolution. Galileo thought that their varying degree of brightness, here and in Galileo's original drawings crudely represented by size, is caused by different refractions of their images through Jupiter's atmosphere. We now know that these moons spin on their axes and in this way bring into view different parts of their surface – parts that reflect sunlight in different degrees.

Galileo also noted, almost in passing, that these moons accompany Jupiter in its twelve-year orbit around the Sun. It is at this point that Galileo, long a secret Copernican, became a public one. Important to this transition and to his subsequent evolution to public defender of Copernicus's cosmology is Galileo's observation "that the revolutions are swifter in those planets which describe smaller circles around Jupiter". Jupiter and its moons validated, by reproducing in miniature, the Copernican cosmos in which those planets closer to the Sun move more swiftly. Furthermore,

> Here we have a fine and elegant argument for quieting the doubts of those who, while accepting with tranquil mind the revolutions of the planets about the Sun in the Copernican system, are mightily disturbed to have the Moon alone revolve about the Earth and accompany it in an annual rotation about the Sun. Some have believed that this structure of the universe should be rejected as impossible. But now we have not just one planet rotating about another while both run through a great orbit around the Sun; our own eyes show us four stars which wander around Jupiter as does the Moon around the Earth, while all together trace out a grand revolution about the Sun in the space of twelve years.

Today, the four brightest moons of Jupiter are named after mythological figures: Ganymede, Callisto, Io and Europa – all conquests of the god Jupiter. Sometimes we, quite appropriately, refer to this group as the *Galilean satellites*.

Altogether the orbits of 67 moons of Jupiter have now been confirmed. Most of these were captured by Jupiter after its formation and, as a consequence, have highly elliptical orbits, highly inclined to the ecliptic. Had Galileo been able to observe some of these smaller satellites with irregular orbits as well as the four brightest ones he might not have seen in Jupiter and its moons a miniature Copernican solar system.

As it was, Galileo became a vocal champion of Copernicus, and as such unwisely responded to detractors who pitted scripture against Copernican cosmology. In particular, these detractors pointed to Joshua's command (Joshua 10:12–13) that the Sun stand still, and to the several Biblical references to the stability of the Earth. Galileo, ever the faithful Catholic, did not doubt that Joshua had miraculously lengthened the day "until the nation took revenge on their enemies", but contextualized the Biblical account as written for readers who believed that the Sun moved around a stationary Earth. Such a defence did not impress Church officials who reserved the office of interpreting scripture to themselves.

19. Kepler's Laws of Planetary Motion (1620)

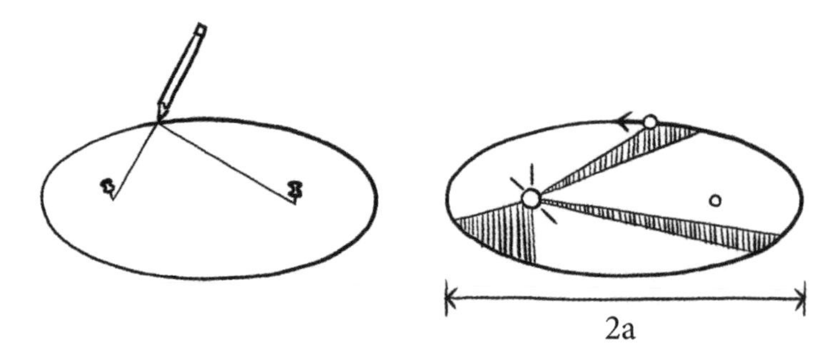

Figure 31

Johannes Kepler (1571–1630) delighted in uncovering the hidden order of the universe. As a young man he embraced the order in Copernicus's heliocentric arrangement of the heavenly bodies. Copernicus's universe was not significantly more accurate than Ptolemy's geocentric one, but it was more ordered, with each of its features logically entailing others.

Crucial to his search for order was Kepler's encounter with Tycho Brahe (1546–1601) in February of 1600. Tycho's personal resources and connections had allowed him to construct the best pre-telescopic astronomical observatories of his time: first the *Uraniborg* observatory on the island of Hven for the Danish king Frederick II and then, after falling out with Frederick's heir, an observatory near Prague financed by the Holy Roman Emperor, Rudolph II. Tycho, more so than any of his predecessors, saw the value of observing the same planet as it progressed through a complete orbit and of determining the uncertainty associated with each observation.

However, Tycho was not a Copernican. Rather, he promoted his own peculiar cosmology in which the five visible planets (Mercury, Venus, Mars, Jupiter and Saturn) moved in concentric circles around the Sun while the Sun itself circled a stationary Earth at the centre of the universe. Tycho wanted Kepler to use his data to verify this system, but because Tycho was uncertain of Kepler's loyalties he shared his data with Kepler with great ambivalence, allowing, for instance, Kepler to view the data but not to copy it for later use.

On Tycho's death in October of 1601 Kepler inherited both Tycho's position as imperial mathematician and Tycho's data on condition that he complete the work of reforming theoretical astronomy on the basis of Tycho's cosmology. Ironically, the high quality of Tycho's data made this task impossible. For none of the established systems, Ptolemaic, Copernican or Tychonic, could incorporate Tycho's observations with the required precision. Eventually Kepler had to abandon these cosmologies, based as they were on combinations of circular motions. Kepler then tried to fit the orbit of Mars to an ellipse with the stationary Sun at one of its two foci – as illustrated in the right-hand diagram of figure 31 – and found that this scheme worked perfectly.

The diagram on the left shows how an ellipse can be constructed with a string, two push-pins, and a pencil. The ends of the string are attached to the push-pins and these, in turn, are stuck in a plane surface. The pencil marks out a circuit in the plane as it slides along and holds the string taut. Hence, an ellipse is the set of points lying in a plane whose distances to each of two points in the plane sum to a constant. The two points are called the *foci* of the ellipse. If the two foci happen to coincide, the ellipse reduces to a circle. A line drawn through the two foci from one end of the ellipse to the other (in the diagram) is twice the *semi-major axis a.*

That each of the planets, Earth included, moves in an ellipse with the Sun at one focus is known as *Kepler's first law of planetary motion*. The diagram on the right-hand side of figure 31 illustrates *Kepler's second law*: A line extending from the Sun to the planet sweeps out equal areas (e.g., the shaded areas) in equal intervals of time. Consequently, the closer a planet approaches the Sun, the faster it moves. While Kepler's first two laws relate the different parts of a single planetary orbit, his *third law of planetary motion* describes a relationship among the orbits of different planets. In particular, the square of the time required for a planet to complete a single revolution around the Sun T is proportional to the cube of the planet's semi-major radius a. In other words, the ratio T^2/a^3 is the same for all of the planets. Kepler suspected that these relationships were the result of the Sun's push and pull on the planet but never discovered the form of that push or pull. He presented evidence for the first two of his three laws of planetary motion in *New Astronomy* (1609) and for the third in *Harmonies of the World* (1619).

Kepler was a generous spirit who sought to conciliate jealous rivals and a faithful Lutheran who twice uprooted his family in order to prevent their forced conversion to Catholicism. He was also an unfortunate who suffered the death of his first wife and eight of his twelve children, the necessity to defend his mother against a charge of witchcraft, and the outbreak of a war, the Thirty Years War, that devastated central Europe. Kepler was deeply pious and deeply grateful for having discovered that for which he had long sought: new evidence of a universal harmony. At the close of his text *Harmonies of the World* he offered this prayer to "Thee, O Lord Creator, who by the light of nature arouse in us a longing for the light of grace".

> If I have been drawn into rashness by the beauty of Thy works, or if I have pursued my own glory among men while engaged in a work intended for Thy Glory, be merciful, be compassionate, and pardon me; and finally deign graciously to effect that these demonstrations give way to Thy Glory and the salvation of souls and nowhere be an obstacle to them.

20. Galileo on Free Fall (1638)

Figure 32

Galileo's fame as a scientist rests on his ability to abstract the essential physics from complicated phenomena, to describe that physics in eloquent words and simple mathematics, and to verify that description with cleverly designed experiments. But Galileo had multiple talents. In fact, Galileo (1564–1642) did so many things well his 20th-century biographer, Stillman Drake, claimed that it is ". . . hard to say whether the qualities of the man of the Renaissance were dominant, or those of our own scientific age". He was an excellent prose stylist, an accomplished visual artist, an ardent gardener, a proficient lute player, and delighted in vigorous debate.

One of Galileo's tactics was to construct "thought experiments" that helped him explore the consequences of a hypothesis – including any absurdities that hypothesis might entail. The diagram above illustrates a thought experiment that Galileo used in *Two New Sciences* (1638). Formally, *Two New Sciences* records a four-day-long conversation among three friends: Salviati, speaking for Galileo; Sagredo, questioning, intelligent, and open-minded; and Simplicio, naively representing what he (Simplicio) understood to be Aristotle's position.

Aristotle (384–322 BCE) had advanced plausible, if superficial, explanations of everyday phenomena. For instance, because objects in motion invariably slow down and come to a stop, continuous motion requires a continuously acting mover. And because heavier objects descend more quickly through water than do lighter ones, heavier objects descend more quickly through all media, including air, in direct proportion to their heaviness or weight and inversely proportional to the resistance of the medium – at least according to Aristotle.

Galileo's persona, Salviati, contests these ideas by arguing as follows. Suppose, in accordance with Aristotle's analysis, a one-pound stone falls with a speed of one cubit per second and a four-pound stone falls with a speed of four cubits per second. If tied together the lighter stone should retard the speed of the heavier one and the heavier one increase the speed of the lighter one and, in this way, result in a speed intermediate between one and four cubits per second. On the other hand, the five-pound package of two stones, considered as a whole, should fall at a speed of five cubits per second. We can avoid this contradiction only if all objects fall from rest at the same speed.

However, Simplicio, the naïve Aristotelian, remained puzzled.

> I am still at sea, he says, because it appears to me that the smaller stone when added to the larger increases its weight and by adding weight I do not see how it can fail to increase its speed or, at least, not to diminish it.

Salviati's response – that is, Galileo's response – to Simplicio surprises us.

> It will not be beyond you when I have once shown you the mistake under which you are labouring. . . . One always feels the pressure upon his shoulders when he prevents the motion of a

> load resting upon him; but if one descends just as rapidly as the load would fall how can it gravitate or press upon him? Do you not see that this would be the same as trying to strike a man with a lance when he is running away from you with a speed which is equal to, or even greater than, that with which you are following him? You must therefore conclude that, during free and natural fall, the small stone does not press upon the larger and consequently does not increase its weight as it does when at rest.

Evidently, when in free fall, neither stone presses upon the other: that is, neither has weight relative to the other – an idea that Albert Einstein exploited to great effect almost three centuries later in constructing his theory of general relativity.

Galileo also appealed to actual experiments. While experiments that could support his case were hard to perform, he knew of at least two that challenged Aristotle's conclusions. Simon Stevin, in 1586, and much earlier John Philoponus (490–570 CE), had dropped objects of very different weight from great heights and found that weight alone makes no significant difference in the rate of fall. (Galileo could also have dropped cannon balls from the Leaning Tower of Pisa for this purpose, but he never claimed to have done so.) It may be that Salviati refers to these earlier experiments in the following passage from the first day of conversation in *Two New Sciences*.

> Aristotle says, "an iron ball of one hundred pounds falling from a height of one hundred cubits reaches the ground before a one-pound ball has fallen a single cubit". I [Salviati] say that they arrive at the same time. You find on making the experiment, that the larger outstrips the smaller by two finger-breaths, that is, when the larger has reached the ground, the other is short of it by two finger-breaths; now you would not hide behind these two fingers the ninety-nine cubits of Aristotle, nor would you mention my small error and at the same time pass over in silence his very large one.

After undermining Aristotle's explanation of falling objects, Galileo proposed that in the absence of a restraining medium all objects accelerate downward at the same rate, in particular, by incrementing their speed equal amounts in equal intervals of time: that is, by about 32 feet (or 9.8 metres) per second every second. Galileo developed the mathematical consequences of this hypothesis and then devised an experiment to look for them. Because objects fall much too quickly for convenient measurement, he rolled spheres down an inclined plane in order to slow down the natural downward acceleration of free fall. This whole procedure (hypothesis, deduction and experimental verification) worked brilliantly for Galileo and since his time has become standard practice for modern physics.

21. Galileo on Projectile Motion (1638)

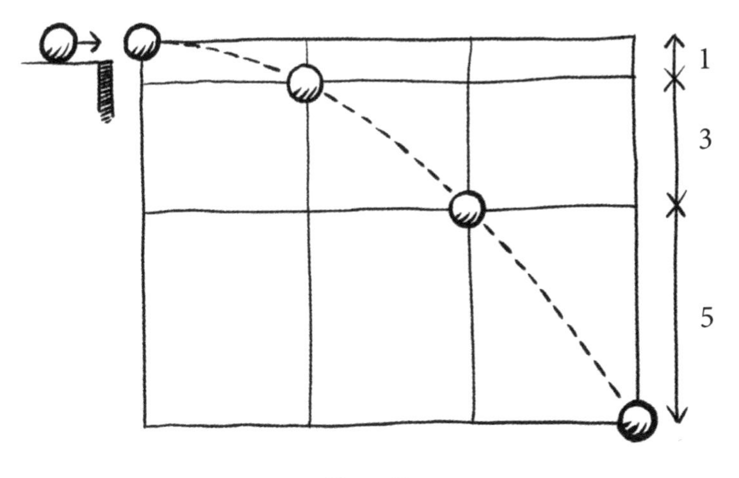

Figure 33

In 1616 the office of the Inquisitor General of the Roman Catholic Church warned Galileo Galilei (1564–1642) to "relinquish altogether the said opinion that the Sun is the centre of the world and immovable and that the Earth moves. . . ." Consequently, Galileo promised not "to hold, teach, or defend in any way whatsoever, verbally or in writing. . . ." the said opinion – a promise that, by publishing *Dialogue Concerning Two Chief World Systems* in 1632, he broke in the most dramatic way. The conceit of the dialogue that placed powerful arguments in favour of a heliocentric universe and weak objections to it in the mouths of fictitious interlocutors fooled no one.

The inquisitors convicted Galileo of "vehement suspicion of heresy" and sentenced him to life in prison – eventually commuted to confinement to his villa in Arcetri. This time Galileo kept the promise extracted from him and no longer spoke or wrote, at least publicly, on the structure of the world. Galileo was humiliated, but we have benefited, for in turning away from cosmology Galileo focused his remaining years on the subject of his greatest achievement: the description of motion. While Galileo's telescopic discoveries of 1610 had demonstrated the unity of earthly and heavenly realms, he nevertheless clung to circular planetary orbits and, to Kepler's dismay, ignored the latter's evidence for elliptical ones. At Arcetri Galileo uncovered the foundations of a new science of motion that to this day remain current – a task for which by 1633 he had been preparing for many years.

The phenomenon of projectile motion had, up to Galileo's time, resisted analysis. Today we might deploy a high-speed, digital video camera and special curve-fitting software, but Galileo had to measure intervals in space with a cord marked off in standard units and time intervals with the outflow of a water clock or, yet more imprecisely, the beating of his pulse.

Galileo first proposed a formal description of projectile motion and then tested the consequences of his proposal against painstakingly obtained experimental evidence. He had, since his school days, been aware of the distinction, originating in the 14th century, between uniform motion, in which a body traverses equal distances in equal intervals of time, and uniformly accelerated motion, in which a body increases its speed by equal amounts in equal intervals of time. Galileo proposed that the motion of a projectile was a combination of these two kinds of motion: uniform motion in the horizontal direction and uniformly accelerated motion in the downward direction. Galileo was aware that the distance traversed by a body moving with uniform speed increases as the first power of the time elapsed, while the distance traversed by a uniformly accelerating body increases as the square of the time elapsed.

Figure 33 shows snapshots, equally spaced in time, of a sphere that has been launched over the edge of a horizontal surface and, as a consequence, continues its uniform motion in the horizontal direction. The distance the object falls in each interval of time, indicated on the right by a sequence of odd integers 1, 3, 5, . . .

ensures that the downward acceleration is uniform. For if during the first interval the sphere falls 1 unit of distance and in the second interval 3 units, in the third 5, and so on, then after 1 interval the sphere has fallen a total of 1 unit of distance, after 2 intervals the sphere has fallen a total of $1+3=2^2$ units, while after 3 intervals $1+3+5=3^2$ units, and so on. In general, after the nth interval of time the sphere has fallen n^2 units of distance. In other words, the sphere falls a distance proportional to the square of the elapsed time – as is characteristic of uniformly accelerating motion. The result of these two motions is a parabolic trajectory, outlined in the diagram, in which the distance fallen is proportional to the square of the horizontal distance traversed.

Such was Galileo's proposal, but how did he confirm it? Heavy objects, whether projected horizontally or simply falling straight down, increase their speed in the downward direction at a rate of 32 feet (9.8 metres) per second every second – accelerating too rapidly for careful observation. So in place of allowing an object to fall freely, Galileo let a bronze sphere roll down an inclined plane and, in this way, slowed down what he called the "natural acceleration" of gravity. He minimized the effect of friction by using smoothly polished, hard wood for his inclined plane. He gathered data by making different measurements on repeated trials of the same experimental arrangement.

Galileo's study of projectile motion, presented in the "fourth day" of his text *Two New Sciences* (1638), broke new ground in ways that prefigure the modern practice of science. He abandoned the Aristotelian idea that continuous motion requires a continuously acting cause, and, indeed, strategically postponed the difficult search for the cause of projectile motion in order to focus on its description. In describing projectile motion, Galileo turned away from a concern with the whole universe and toward a phenomenon that could be isolated from its environment and from which he could abstract its essential elements. He searched for simple mathematical relationships that described this physics and tested these relationships with experiments that reproduced the idealized situation as closely as possible.

Galileo's analysis of projectile motion is the basis of a simple device often used by physics teachers. Two identical steel balls are mounted on a wooden block. A spring-loaded lever launches one of the balls in the horizontal direction while simultaneously releasing the other and allowing it to drop from rest. The two steel balls strike the floor at the same time with a satisfyingly single "plunk". Thus, projectiles and freely falling objects accelerate downward at the same rate: 32 feet or 9.8 metres per second every second.

22. Scaling and Similitude (1638)

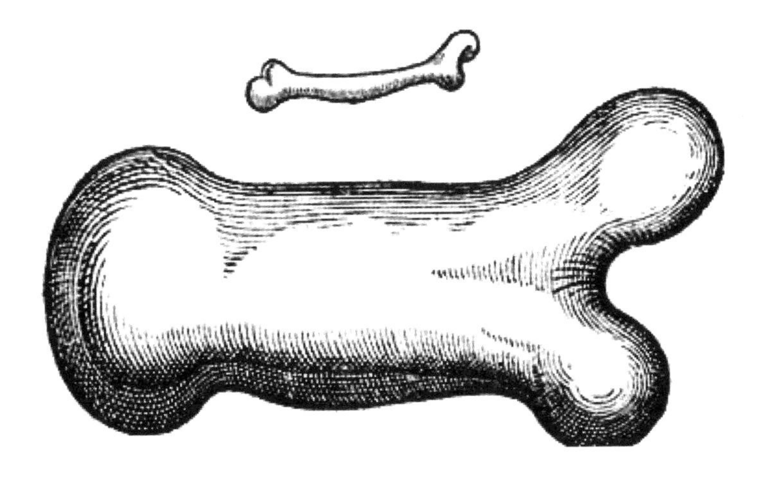

Figure 34

This drawing, in Galileo's own hand, appears in his *Two New Sciences* (1638). Its exquisite shading tells us that these are not mere shapes for which outlines would have sufficed. Galileo wants us to see these figures as *bones* – bones with which he can illustrate the concepts of *scaling* and *similitude* and take the first steps toward a theory of appropriate size.

Mathematical objects with the same shape, such as two triangles that differ only in scale, are said to be *similar*. Three-dimensional objects, for example, pyramids, with parts in the same proportions yet of different size, are also similar. However, most natural objects and animals with similar shapes occupy a more or less limited range of sizes within which they can be large or small versions of themselves.

Sometimes our imaginations run away with the idea of scaling. In Jonathan Swift's *Gulliver's Travels* a storm washes Lemuel Gulliver onto the island of Lilliput. Lilliput's inhabitants are twelve times smaller than Gulliver. Later Gulliver is marooned on the island of Brobdingnag whose inhabitants are twelve times larger. Otherwise these relatively small and large people are, in virtue, in vice, in wisdom, and in folly, much like the humans of Swift's time.

As entertaining or as instructive as such tales may be, Galileo would have recognized the flaw in Swift's descriptions – for it is one thing to imagine indefinitely large or small mathematical objects and quite another to imagine such objects clothed with physical attributes within natural environments. According to Galileo, a giant twelve times as large as a normal human being would collapse under his own weight.

Galileo reasonably supposed that the cross-section of a limb determines the limb's strength. After all, an animal's muscles push or pull across a cross-sectional area. Furthermore, a bone is like a wooden beam. When a beam breaks it does not break everywhere at once but breaks through in a single jagged cross-section. Yet the weight that a bone or beam must support is directly proportional to the mass of the whole structure to which it belongs. In general, the strength of an animal, or of any structure, is directly proportional to the area of its cross section while the weight it supports is directly proportional to its total volume.

Since cross-sectional area increases with the square of a scale factor L, that is, with L^2, and volume increases with the cube of this scale factor, that is, with L^3, the ratio of an object's strength to the weight it supports varies as L^2/L^3, that is, as $1/L$. This dependence has become known as the *square-cube law*. By reason of the square-cube law, the larger an animal or structure, the less able it is to support its own weight.

For the same reason, smaller creatures of the same general shape have a relative advantage in strength. "Thus," says Galileo,

> A small dog could probably carry on his back two or three dogs of his own size; but I believe that a horse could not carry even one of his own size.

He could also have noted that an ant can carry more than ten times its own weight.

The larger of the two bones in Galileo's drawing is about 10 times wider than the smaller one but only about 3 times as long. Therefore, these bones are *not* geometrically similar. Instead Galileo chose these proportions in order to better preserve their strength to weight ratio. Since the cross-sectional area of the larger bone is 10^2 or 100 times larger than that of the smaller bone, the larger bone can carry or support 100 times more weight than that carried or supported by the smaller bone. And if the animal to which the larger bone belongs is relatively thick-limbed and squat, that is, as they say "overbuilt" as shown here, it may manage to weigh only about 300 times more than the animal to which the smaller bone belongs. Thus, the larger animal would be relatively weaker than the smaller animal but still within the realm of possibility. If, on the other hand, the larger animal were geometrically similar and 10 times larger than the smaller one, it would weigh 1000 times more while being only 100 times as strong – probably not within the realm of possibility. Apparently, nature departs from *geometric similitude* in order to preserve relative strength.

The science of determining what ratios are important in what contexts is called *dimensional analysis*. Comparative zoologists as well as engineers who build and test scale models, say, in wind tunnels and towing tanks, seek out, with the help of dimensional analysis, these ratios. In most cases the structure's interaction with its environment is crucial to their task. Small water-borne insects, for instance, must cope not so much with gravity but rather with the surface tension and viscosity of water. And small mammals develop special strategies for maintaining their body temperature. Preserving the ratio of relevant forces (*dynamical similitude*), of relevant velocities (*kinetic similitude*), and of relevant thermal quantities (*thermal similitude*) is usually more important than preserving geometric similarity when scaling up or down.

23. The Weight of Air (1644)

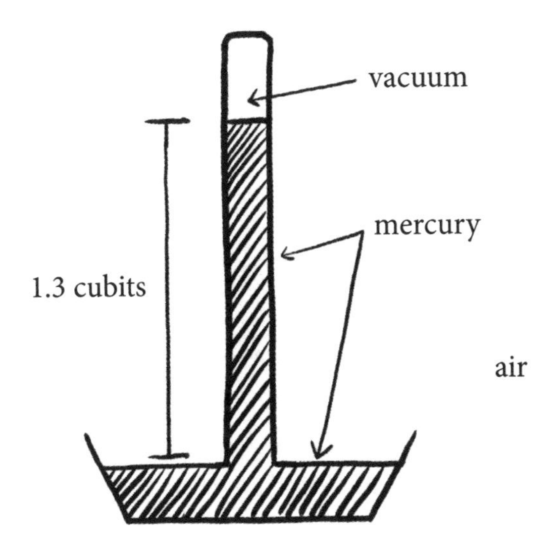

Figure 35

The element mercury is relatively rare, but it does not blend easily with other elements in the Earth's crust, and, for this reason, is sometimes found isolated in mineral deposits. These deposits have been mined for millennia for, in spite of mercury's toxicity, our ancestors valued it, a shiny liquid metal, as a medicine, as an ornament and for its high density.

Evangelista Torricelli (1608–1647) put mercury's density, roughly 14 times that of water, and its liquidity to good use in devising the first *barometer* as shown in figure 35. He prepared a narrow glass tube sealed at one end, filled it with mercury, stopped the open end with his finger, inverted it, and inserted it into a basin of mercury. Upon removing his finger the mercury in the tube fell and left a column about 1.3 cubits (2.5 feet, 29 inches, or 74 centimetres) high.

In this way Torricelli created something at the top of the tube that had long been thought impossible: a region containing nothing – that is, a vacuum, a *Torricellian vacuum.*

But is it not true that "Nature abhors a vacuum?" A few years earlier Galileo (1564–1642) had been drawn to reflect on this abhorrence by the testimony of a workman who had been called upon to repair a suction pump that drew water from a well (illustrated in figure 36). According to Galileo's text *Two New Sciences* (on the first "day" of conversation) the workman claimed that

> the defect was not in the pump but in the water which had fallen too low to be raised through such a height; and he [the workman] added that it was not possible, either by a pump or by any other machine working on the principle of attraction to lift water a hair's breadth above 18 cubits; whether the pump be large or small this is the extreme limit of the lift.

Galileo imagined that the vacuum created at the top of the pump suspended the column of water and that if the column got too long the water would break under its own weight just as a rod of wood or iron suspended at its top would, if sufficiently long, also break.

Torricelli respected Galileo to the point that he waited upon the man, blind and under house arrest, during the last three months of his life. However, Torricelli dismissed Galileo's hypothesis that the vacuum itself suspends the weight and held that, on the contrary, because "we live submerged at the bottom of an ocean of air", the surrounding air, by pushing down on the surface of the barometer basin and the well water, pushes up the columns of mercury and water. Thus, air will support a column of mercury, about 1.3 cubits (2.5 feet, 29 inches, or 74 centimetres) high, and will similarly support a column of water, about 18 cubits (34 feet or 10 metres) high – just so much higher than the density of mercury is greater than the density of water, roughly a factor of 14. Torricelli intended that his barometer measure weather-related changes in the weight of

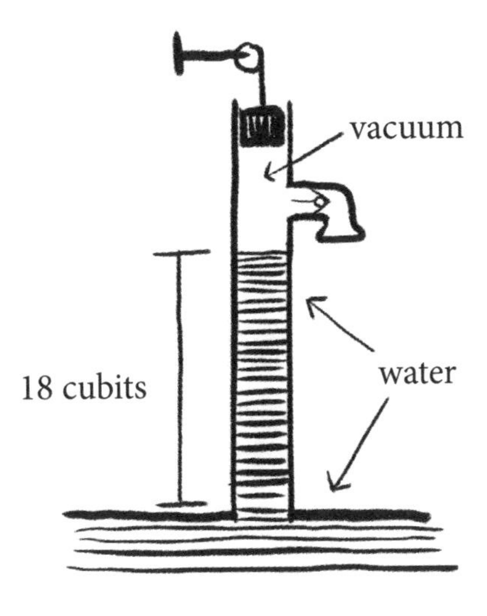

Figure 36

air but this effect could not, in his time, be easily separated from temperature-induced changes in the volumes of its glass and mercury parts.

Torricelli's explanation survived while Galileo's did not. For if the barometer's column of mercury was pushed up by the weight of the surrounding air rather than held up by the abhorrence of the vacuum, the mercury column should fall as the barometer is carried to a higher elevation where the air was known to be thinner and less weighty. The philosopher, mathematician, and physicist Blaise Pascal (1623–1662) tested this idea by arranging for his sister's husband, the judge Florin Périer, to transport a barometer up a mountain, the so-called Puy de Dôme, that rises 900 metres (3,000 feet) above the nearby municipality of Clermont-Ferrand in France.

Pascal's 1648 report *The Great Experiment on the Weight of the*

Mass of the Air includes his careful instructions to his brother-in-law and Périer's exciting description of the actual experiment. Its result: a complete vindication of the idea that air has weight. Périer constructed two mercury barometers and left one attended at the foot of the Puy de Dôme while he and his companions carried the other to its top. There he found that the mercury column had fallen 8 centimetres (3 $^{1}/_{6}$ inches) during the course of their ascent. Périer repeated this test on a smaller scale by ascending the 37-metre (120-foot) high tower of the Notre Dame de Clermont as did Pascal himself by ascending a 46-metre (150-foot) high tower in Paris. In each case the mercury column fell a distance proportional to the ascent of the barometer. The only credible explanation was Torricelli's that "we live submerged at the bottom of an ocean of air". Pascal concluded his report with these words:

> Does nature abhor a vacuum more in the highlands than in the lowlands? . . . Is not its abhorrence the same on a steeple, in an attic, and in the yard? . . . let them [the Aristotelians] learn that experiment is the true master that one must follow in Physics; that the experiment made on mountains has overthrown the universal belief in nature's abhorrence of a vacuum, and given the world the knowledge, never more to be lost, that nature has no abhorrence of a vacuum, nor does anything to avoid it; and that the weight of the mass of the air is the cause of all the effects hitherto ascribed to that imaginary cause.

24. Boyle's Law (1662)

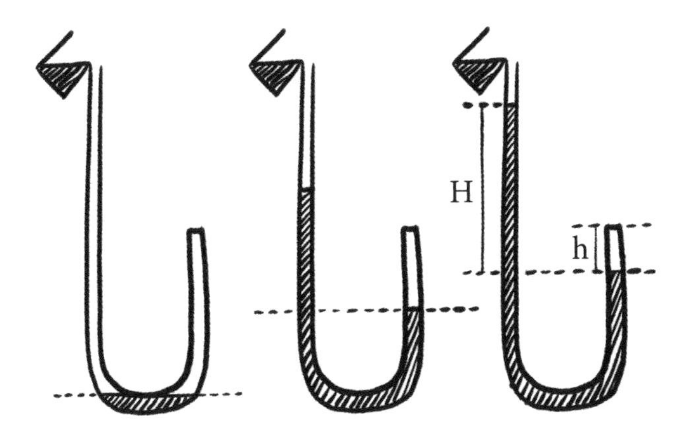

Figure 37

Robert Boyle's (1627–1691) first trial of the experiment depicted here ended in disaster. He had a long glass tube bent in the shape of a "U" with its two unequal legs parallel to one another. The longer leg was more than six feet in length and the shorter one was sealed at its end. Then he poured mercury into the open end of the long leg of the tube. His object was to record paired values of the distances marked H and h in the third panel of figure 37, H indicating how much higher the mercury is in the longer leg than in the shorter one and h indicating the length of the column of air trapped in the shorter leg. But before he could gather data he accidently broke the unwieldy tube and, presumably, spilled the expensive mercury.

As the well-educated son of the fabulously wealthy first Earl of Cork, Boyle had both the know-how and the means to do this experiment – and to redo it as necessary. And because Boyle was acquainted with Torricelli's barometer (1643) and with Pascal's demonstration of the weight of air (1648) he was able to interpret his results as a demonstration of what is known, in England and in the United States, as *Boyle's law*, the first enunciation of which Boyle published in an appendix (1662) to an earlier work, *A Defence of the Doctrine Touching the Spring and Weight of Air* (1660).

Figure 37 illustrates the idea behind Boyle's law. In the first panel the air in the shorter leg is just barely connected to the air in the longer leg which, in turn, is open to the atmosphere. Given the behavior of Torricelli's barometer, we know that the pressure of the air that surrounds us is enough to hold up a column of mercury a little more than 29 inches high or, alternatively, a column of water 34 feet high or a column of air extending to the top of the atmosphere. As the mercury continues to flow into the longer leg, the air in the shorter leg is cut off from the atmosphere. And as the mercury in both legs rises – more quickly in the longer leg than in the shorter one – the air in the shorter column is compressed. Boyle kept a record of the co-ordinate values of the three quantities H, $H+29$, and h to the nearest 16th of an inch. The numbers below are Boyle's values listed to the nearest inch.

When the mercury in the longer leg is 29 inches higher than in the shorter one, the air pressure in the shorter leg must be enough to hold up the 29-inch high column of mercury and a column of air extending to the top of the atmosphere (equal in weight to another 29-inch high column of mercury). At this point the column of air in the shorter leg is compressed from its initial value of 12 inches to 6 inches in length. As more mercury is poured into the longer leg the quantity $H+29$ grows and the quantity h shrinks in inverse relation so that $(H+29) \propto 1/h$. Thus, when $H=2\times29$ and, therefore, $H+29=3\times29$, the original value of h is diminished by a factor of 3. Since $H+29$ is directly proportional to the pressure P exerted on and by the column of trapped air and h is proportional to the trapped air's volume V, these data demonstrate Boyle's law, which in algebraic form is $P \propto (1/V)$ as illustrated in figure 38.

H	H+29	h
0	29	12
1	30	12
3	32	11
4	33	11
6	35	10
8	37	10
10	39	9
12	41	9
15	44	8
18	47	8
21	50	7
25	54	7
29	58	6
35	64	6
41	70	5
49	78	5
58	87	4
71	100	4
88	117	3

Boyle may have known that $P \propto (1/V)$ obtains only when the temperature of the gas is constant – as it was in his experiment – but never mentioned this limiting condition. It was Edme Mariotte (1620–1684) who, having independently discovered Boyle's law in 1667, made this condition explicit. For this reason Europeans often refer to *Mariotte's law* or to the *Mariotte-Boyle law*.

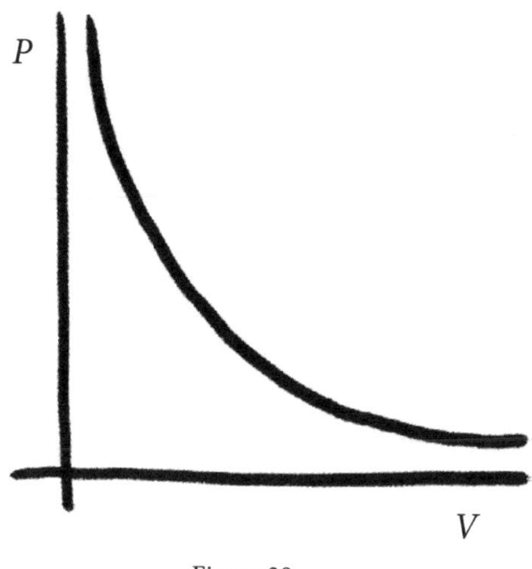

Figure 38

Boyle, like Francis Bacon (1561–1626) before him, was a champion of empirical study – such as that which led to his eponymous law. Also, like Bacon he was suspicious of overarching theories. Speculative hypotheses and mathematical expression of physical laws were helpful only when motivated by data gathered from close observation and careful experimentation.

Like many from wealthy Anglo-Irish families Boyle was educated partly at home with private tutors and partly in an English "public" school: in this case, Eton College near Windsor. He left England for study on the Continent with a tutor as a 12-year-old in 1639. While in Europe Boyle's father died and left Robert a rich young man. When he returned in 1644 the Irish were rebelling against English rule and England was in the midst of a civil war. His brothers and sisters were on both sides of this latter conflict but Boyle became a partisan of neither. As he remarked in a letter to his former tutor, he felt exposed "to the injuries of both parties, and the protection of neither". Throughout his life Boyle observed "a very great caution" in all matters political and religious.

Boyle is sometimes called the "father of chemistry" – probably for rejecting both the Aristotelian doctrine of four elements (earth, air, fire and water) and the Paracelsian doctrine of three principles (salt, sulphur and mercury) and in their place emphasizing that all chemical phenomena should be understood in terms of the mechanics of particles in motion. During the late 1640s while in London he began meeting weekly with a group of like-minded natural philosophers to witness and discuss physical demonstrations. He called this group, which later evolved into the Royal Society of London, "the invisible college".

Boyle had wide interests and wrote prolifically on medicine, theology and language as well as on physical science. Nearly half of his literary output was devoted to theological topics, especially to the relation of theology to the new philosophy of experimental science. Boyle's last will and testament established a series of lectures on Christian apologetics that recently (2004) have been revived with the express purpose of exploring the relationship between Christianity and science.

25. Newton's Theory of Colour (1666)

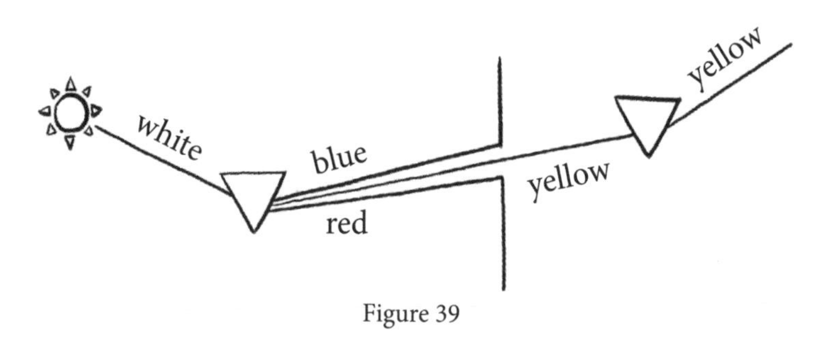

Figure 39

Isaac Newton (1642–1727) entered Trinity College, Cambridge University in 1661, the year after the Restoration of the English monarchy that, in turn, followed the beheading of Charles I and the decade-long dictatorship of Oliver Cromwell. At that time the university had entered a long period of decline from which it did not emerge until after Newton's death. Although nominally dedicated to educating young men, chiefly for the clergy, the fellows of Cambridge were not obliged to tutor, to lecture, to publish, or even to remain in residence. Many, in fact, chose to absent themselves for months and years at a time. Even so they drew their stipends. Only three offences were cause for dismissing a fellow: voluntary manslaughter, becoming a heretic, and getting married.

Yet in many ways Cambridge was a perfect place for Newton. He was self-motivated and independent-minded and would not have followed the guidance of a good teacher were one available. All he needed were some books and tools (he installed a lathe in his student lodgings), and to be left alone. He learned by continually thinking on a subject of his own choosing. His youthful obsessions were mathematics, mechanics and optics.

Newton became, in turn, Master of Arts, Fellow of Trinity

College, and, at the age of thirty, the Lucasian Professor of Mathematics, one of the most lucrative professorships in the kingdom. So steady was his advancement that Richard Westfall, his 20th-century biographer, believed that, in an age when the Church or the Court dictated most academic appointments, the young Newton must have had a powerful patron – now unknown to us. Upon assuming his professorship Newton was obliged to give a series of inaugural lectures. The phenomenon of colour was the subject of these lectures.

Newton's theory of colours is so thoroughly our own, we struggle to imagine another. But for two thousand years people believed that sunlight was pure and simple. Colour was somehow added to originally colourless sunlight during its reflection from or refraction (or bending) through different transparent materials. Newton, for instance, tried to imagine light as a collection of identical globules that, like tennis balls, could acquire a spin during reflection and refraction. Different rates of spin would correspond to different colours. But these ideas fell by the wayside once he bought some prisms and started experimenting on his own.

Newton's experiments suggested a radically different theory of colour. He arranged for sunlight to enter his chamber through a small circular hole in a closed shutter and to fall upon a triangular glass prism as shown on the left side of figure 39. That such prisms produce coloured light was already so well known as to be *celebrated*, to use Newton's word. René Descartes, Robert Boyle and Robert Hooke (1635–1703) had all reported on the phenomenon, but none had projected a refracted beam on a surface more than a few feet distant from the prism that caused the refraction. As a result the size and shape of their refracted beams were not noticeably changed. Newton, on the other hand, projected his beam onto the opposing wall of his chamber 22 feet distant and found that the originally circular image now extended five times more in one direction than in the other and displayed a series of colours in the extended direction. Moreover, the size of the oblong image increased in direct proportion to the distance from the prism. Newton concluded that sunlight is a composition of different colours, and the different colours refract or bend in different amounts, blue more than yellow and yellow more than red.

Figure 39 also shows an extension of this experiment that Newton used to confirm these ideas – what he called his *experimentium crucis*. As before, a beam of sunlight refracts through a triangular glass prism and the oblong image projects on an opaque surface. This time the light of a single colour (here yellow) passes through a hole in that surface and falls on a second prism. If a prism creates colours by modifying the beam rather than by separating the beam into its constituent colours, the second prism should also modify the colour of the yellow beam. But that beam remained yellow on passing through the second prism. Newton still believed that light was composed of particles, but he was now also convinced that (1) sunlight is a heterogeneous mixture of colours, (2) which refract in different amounts, and (3) upon refraction separate into a continuous spectrum of colours. Furthermore, (4) opaque objects appear differently coloured because they preferentially reflect one colour.

Newton's theory of colour had an immediate consequence. He abandoned his effort to produce lenses, for refracting telescopes, that perfectly focus starlight. For since lenses focus light by refracting – that is, by bending their rays – and different-coloured rays bend in different amounts and starlight, like sunlight, is a mixture of different colours, starlight will never focus to a single point in a refracting telescope as is required for perfect image formation. This realization may have motivated Newton to build the first *reflecting telescope* in 1668 that avoids refraction altogether.

Richard Westfall confesses, in the preface to his 900-page biography of Newton, *Never at Rest* (1983), that the more he learned of the man, the more alien he seemed. Newton never married and, apparently, had few friends. Yet his Cambridge colleagues viewed him with respect if not awe. Newton would sometimes draw figures, such as the one in figure 39, in the newly prepared gravel walks of Trinity College, and they, for a time, would carefully walk around these drawings in order to preserve them for the Lucasian Professor's use.

26. Free-Body Diagrams (1687)

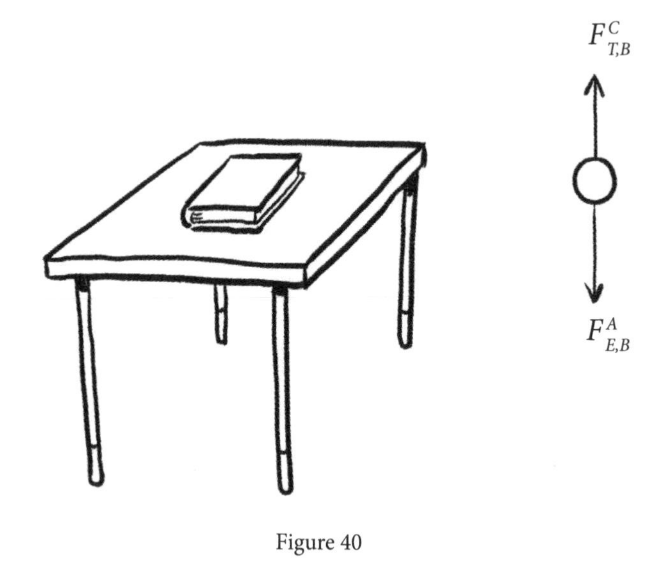

Figure 40

At a certain stage in his or her study of physics the student is called upon to master an important diagrammatic tool: the free-body diagram. Free-body diagrams help us analyse a situation in terms of forces. A general Newtonian principle is that only bodies exert forces on one another. What are the forces and what are the bodies that exert and that experience these forces? A free-body diagram helps us keep track of our answers.

Consider, for example, the seemingly simple situation of a book resting on a table that in turn rests on the ground. We know that both the book and the table are heavy: that is, both are attracted downward toward the centre of the Earth. This gravitational force – after the Latin *gravis* for *heavy* – is an example of an *action-at-a-distance force* because the bulk of the Earth need not touch either the book or the table in order to pull them down. In contrast, an object that must touch another object in order to exert a force on it is called a *contact force.*

Figure 40 shows the book and the table and a free-body diagram of the book. Free-body diagrams represent objects as dots or circles and the forces on the object as directed line segments or arrows whose tails are attached to the object and whose heads point in the direction of the force. The length of the arrow is proportional to the magnitude of the force applied to the object. A longer arrow represents a larger-magnitude force and equal-length arrows represent equal-magnitude forces. We also attach superscripts to each force symbol indicating the nature of the force – action-at-a-distance or contact – and subscripts indicating the force's source and the object to which it is applied. Thus, $F^{A}_{E,B}$ stands for the action-at-a-distance force exerted by the Earth on the book. We usually call this particular force the book's *weight.*

If the force of gravity were the only force on the book, it would, according to Newton's second law, accelerate downward. However, our initial description indicates that the book is at rest. Therefore, its acceleration must be zero. Consequently, the net force on the book must vanish and a force other than the Earth's gravity must be applied to the book in order that it remain at rest.

We deduce the source of this additional force on the book by a process of elimination. A force must be either an action-at-a-distance force or a contact force. The gravitational pull of the Earth is the only action-at-a-distance force on the book. (We ignore the relatively small gravitational attraction between book and table.) And, clearly, since the table is the only object touching the book, the table is the only object capable of exerting a contact force on the book. Therefore, because, and only because, we know the book is at rest, we know that the contact force the table exerts on the book $F^{C}_{T,B}$ points in the opposite direction and is of equal magnitude to the action-at-a-distance force the Earth exerts on the book $F^{A}_{E,B}$.

But how do contact forces work? How, in particular, does the table exert a force on the book? Interestingly, the surfaces of the table and the book behave approximately as if they were composed of a multitude of tiny springs that resist compression. When the book is placed on the table, a book-shaped array of "springs" in the table is compressed until the total upward force of the table on the book is equal and opposite to the downward force of gravity on the book. The depression in the table is typically so slight as to be invisible to the naked eye. If the table were not strong enough to generate and maintain this contact force, the book would crash through the table.

A similar kind of analysis applies to the table, with the interesting difference that now two objects, the ground on which the table rests as well as the book, touch and push on the table. For this reason the corresponding free-body diagram of the table in figure 41 shows three forces on the table: the downward gravitational force of the Earth on the table $F^{A}_{E,T}$, the downward contact force of the book on the table $F^{C}_{B,T}$, and the upward contact force of the ground on the table $F^{C}_{G,T}$. These forces must sum to zero because we know the table is at rest and, therefore, not accelerating.

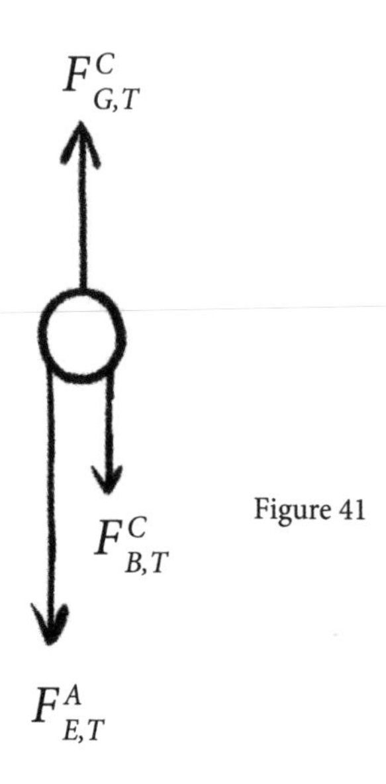

Figure 41

Free-body diagrams can also be drawn for bodies that are not at rest. Consider, for instance, a pear that has fallen from its tree but has not landed on the ground. Figure 42 illustrates the falling pear and its free body diagram. Since nothing touches the pear, the gravitational force of the Earth on the pear $F^A_{E,P}$ is the only force on the pear. Thus, the net force on the pear is its weight $F^A_{E,P}$. According to Newton's second law, the pear will accelerate toward the centre of the Earth at a rate equal to $F^A_{E,P}/m$ where m is the mass of the pear.

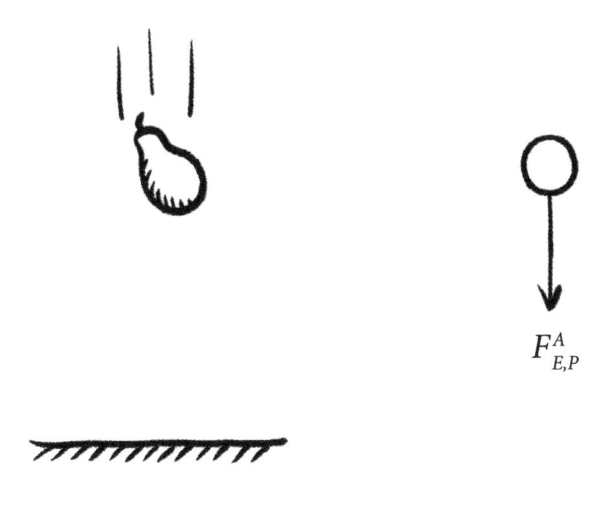

Figure 42

27. Newton's Cradle (1687)

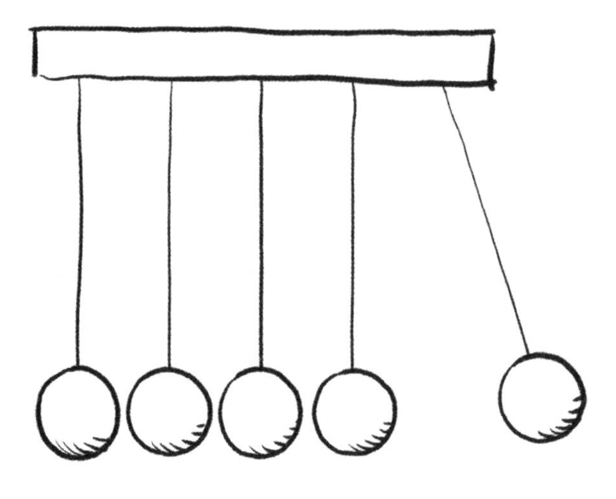

Figure 43

Many of us have played with a "Newton's cradle", composed of several steel balls suspended at rest, as illustrated in figure 43. Our concern, however, is best illustrated (in figure 44) with a simplified version of this toy composed of only two steel balls. The left-hand panel of figure 44 below shows the black ball hanging at rest while the elevated white ball is released and allowed to swing down and strike the black ball. The right-hand panel shows the two balls shortly after their collision. The white ball is now at rest and the black ball has completely captured the motion originally in the white ball. Eventually the black ball will swing up to a level close to that originally occupied by the white ball. Sometimes we hear this device referred to as a "double pendulum", so called because two pendula mimic the motion of one.

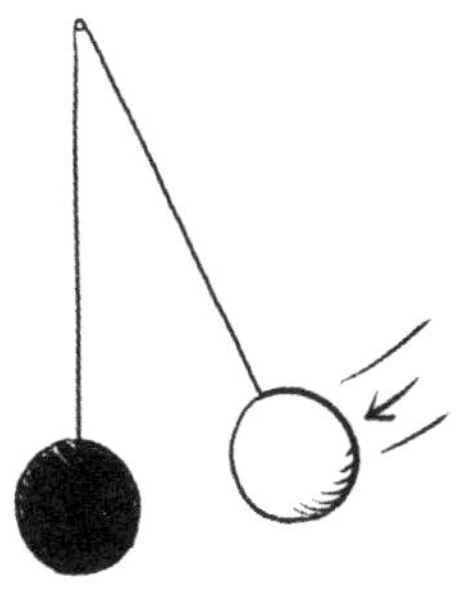
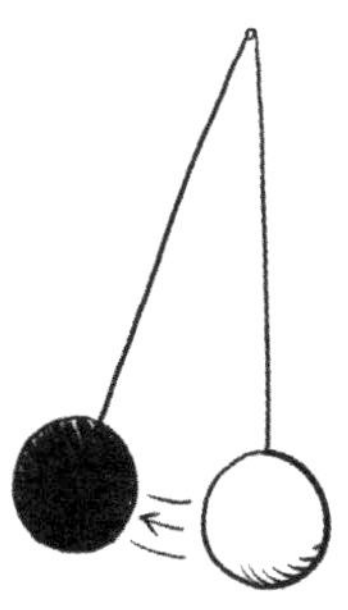

Figure 44

The collision of these two steel balls is approximately *elastic*. In a perfectly elastic, head-to-head collision of two identical objects, one originally at rest and the other moving, the one originally at rest completely captures the motion of the originally moving object. The collision of two billiard balls is approximately elastic. However, collisions are not always perfectly or even approximately elastic. One could, for instance, easily arrange for the two swinging balls to stick together on impact and so reproduce an example of a perfectly *inelastic collision*. After a perfectly inelastic collision identical balls move off together at half the speed of the moving ball just before collision.

Could there be one principle behind both, indeed behind all, kinds of collisions? Newton found such a principle. It was his effort to understand collisions that set him on a path that led to his three laws of motion, which encapsulate the foundations of classical mechanics.

Observe that in both kinds of collision the ball originally at rest speeds up by the same amount that the ball originally in motion slows down. The technical word for speeding up and slowing down is *acceleration*; speeding up is positive acceleration and slowing down is negative acceleration. Therefore, both when the two identical balls stick together and when they do not, the ball originally at rest accelerates by the same amount that the ball originally in motion decelerates. Given that, according to Newton's second law, forces cause acceleration, the force on one of the balls must be equal and oppositely directed to the force on the other ball.

The general principle behind this last statement is *Newton's third law*. Newton's third law is probably the least understood of Newton's three laws of motion – possibly because its original Latin was first rendered into English by the seemingly meaningless phrase: *For every action there is always an equal and opposite reaction*. This widely reproduced statement is more a mnemonic for than an accurate description of the law. When used correctly this mnemonic should remind us that all forces occur in action-reaction pairs of equal magnitude and opposite direction. Thus, whenever object A exerts a force on object B, object B exerts a force of equal magnitude and opposite direction on object A.

The first part of figure 45 again shows colliding balls, while the second part consists of two free-body diagrams that label the forces on each ball. These free-body diagrams clearly show that the left ball exerts a contact force on the right ball $F^C_{L,R}$ equal in magnitude and opposite in direction to the contact force the right ball exerts on the left ball $F^C_{R,L}$. These diagrams are typical in another way. They show that each of the two forces in an action-reaction pair applies to a different object.

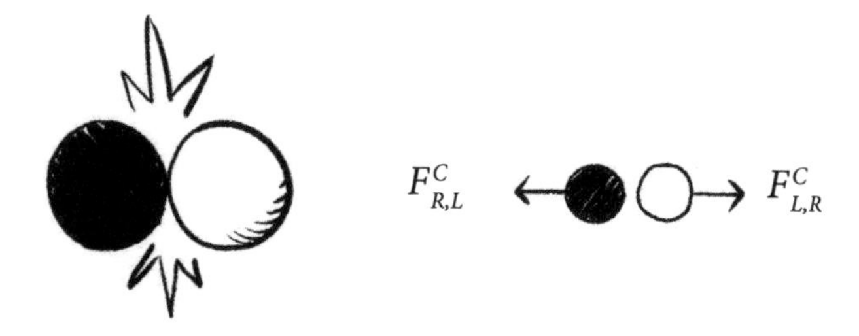

Figure 45

We take advantage of Newton's third law every day. For when beginning to walk from a resting position our body accelerates in the forward direction, and, according to Newton's second law, whenever a body accelerates there is a net force on that body in the direction of its acceleration. What is this force that causes us to begin walking? The force of gravity is in the wrong direction. It could pull us through the floor, but cannot accelerate us along the floor. We cleverly solve this physics problem by pushing backward on the floor with one foot. Consequently, the floor must, according to Newton's third law, exert a force on this foot that accelerates us in the forward direction. Of course, this happens whether or not we are aware of Newton's third law. Our bodies understand Newton's third law and exploit it every day.

28. Newtonian Trajectories (1687)

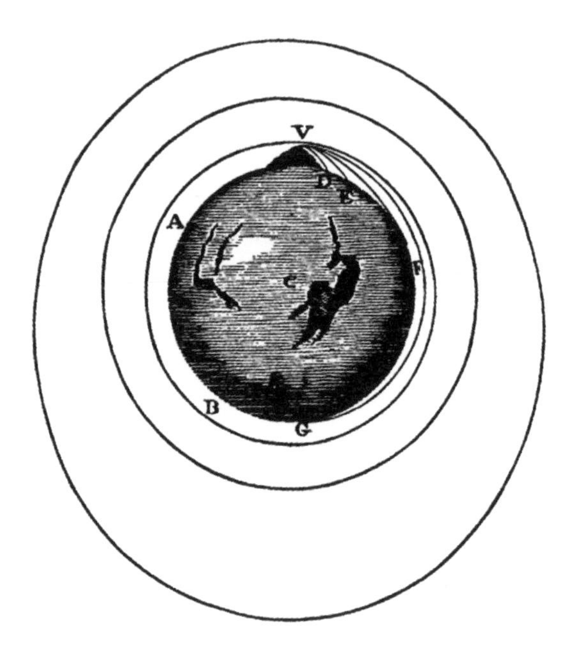

Figure 46

Galileo's telescopic observations of the mountains and valleys on the Moon suggest that the earthly and heavenly realms are similarly composed. Newton (1642–1727) confirmed this suggestion by generalizing to cosmic scale another of Galileo's discoveries: the parabolic trajectory. Pitch a baseball horizontally and imagine its trajectory unfolding in space and time. The baseball covers ground in direct proportion to the first power t of the time elapsed and falls downward in direct proportion to the second power t^2 of the time elapsed. The result is a parabolic trajectory.

But the Earth's surface is not flat; neither is the strength of its gravitational attraction independent of the distance from its centre. Things do not really fall *down*, but rather *toward the centre*, and the magnitude of the resulting *centripetal* (to the centre) *acceleration* diminishes with distance from the centre. What is unimportant in the context of a pitched baseball becomes important for trajectories that are as large as the Earth itself. This is the lesson of figure 46, taken from Newton's major work *Philosophiae Naturalis Principia Mathematica* (1687).

Newton's *Principia* completes a line of thought he had begun in 1664–66, during which time Cambridge University had closed and sent its students and fellows away in order that they might protect themselves from an epidemic of the plague then sweeping through England's cities and towns. Newton returned to his boyhood home in the village of Woolsthorpe and to his twice-widowed mother Hannah Smith. He built bookshelves, read, and thought – and tried to do little else. He invented a scheme of calculation in which ratios of indefinitely small quantities and their infinite sums had finite limits – a scheme we now call *the calculus.*

He also reflected on the effect of gravity reaching far beyond his mother's apple-laden trees all the way up to the orbit of the Moon. How, he asked, would the Moon fall if it fell like an apple? Newton supposed the Moon's orbit was the result of a double tendency: one to continue in straight-line motion and the other to accelerate toward the centre of the Earth. He found that this combination produced, in the simplest case, a circular orbit. Newton also wondered how the magnitude of the acceleration caused by gravity diminished with distance from the Earth's centre – as the first power of the inverse distance $1/d$, or as the second power $1/d^2$? Guided by Kepler's third law – *The periods of the planetary orbits increase as the* 3/2 *power of their average distance from the Sun* – he settled on the second power. His calculations were "pretty nearly" consistent with what he observed: a month-long revolution of the Moon around the Earth.

The 17th-century equivalent of a pitched baseball is a cannonball fired from a horizontally directed muzzle. Place the cannon on top of the highest mountain on the Earth, eliminate the Earth's atmosphere, and you have the situation illustrated in Newton's drawing. At relatively low velocities the cannonball follows what appears to be a parabolic trajectory. As the cannonball's initial speed increases, its range increases until the cannonball orbits the Earth in a perfect circle. The two outermost orbits are ellipses with one focus at the centre of the Earth. The diagram suggests that a parabolic trajectory is, in actual fact, a small part of an ellipse. The larger part of the elliptical trajectory of a baseball or a cannonball is never realized because the bulk of the Earth intervenes.

The *Principia* demonstrates mathematically what the drawing can only suggest. The inverse square law of gravitation and Newton's laws of motion result in the approximate parabolic trajectories of projectiles near the surface of the Earth, in the circular and elliptical orbits of artificial satellites and the Moon, and in the elliptical and hyperbolic orbits of planets and comets.

The *Principia* proposes a universally operating law of gravitation according to which each point of mass in the universe attracts every other point of mass with a force proportional to the product of the masses and the inverse square of their separation. Newton carried his calculations so far as to develop an algorithm for predicting the size and timing of high tides by accounting for the oceans' simultaneous three-fold attraction to Earth, Moon, and Sun.

Universal gravitation was a success, but Newton knew that the idea was incomplete. After all, how could the Sun exert a force that kept the Earth in its orbit without the benefit of an intervening mechanism? For Newton and other 17th-century natural philosophers, to explain a phenomenon meant to reveal the mechanism in terms of which objects, in direct contact, push or pull on one another. Newton abhorred the concept of a force that could project itself across empty space. Yet he cautiously refrained from offering a mechanical explanation of gravity, and, in the end, accepted the universal law of gravitation as a means for efficient and precise mathematical description.

The first (1687) edition of the *Principia* made Newton famous. During his remaining 40 years Newton revised and extended the *Principia* and, at long last, brought his optical and mathematical writings into print. In 1696 he left his professorship at Cambridge and accepted an appointment as Warden of the Mint and, in 1700, as its Master. He represented Cambridge in Parliament (1701) and was elected President of the Royal Society (1703). Queen Anne knighted him *Sir Isaac Newton* in 1705.

Newton died in 1727 but his intellectual legacy endures. In no sense has Newton's physics been overthrown. Relativity and quantum mechanics embed Newtonian physics as a limiting case – the particularly important case that describes the world of human-sized objects. The Newtonian synthesis remains the core of physics education and provides the structure that grants meaning to new discoveries.

29. Huygens's Principle (1690)

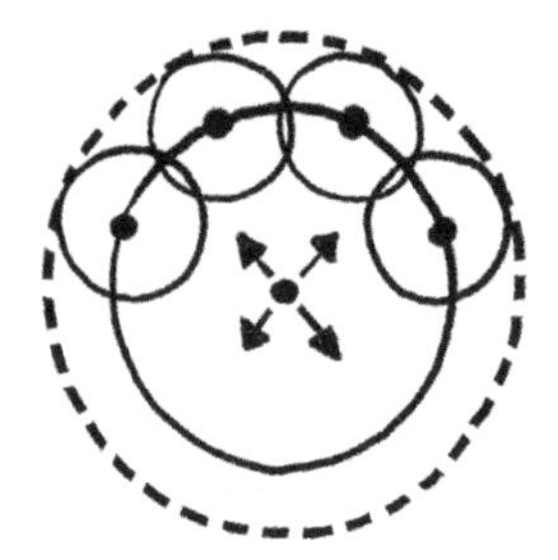
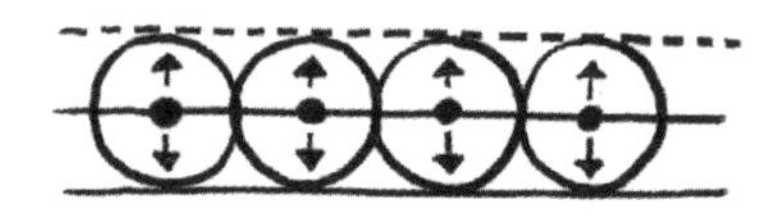

Figure 47

Constantijn Huygens intended that his son Christiaan (1629–1695) follow him into the Dutch diplomatic service and for this reason gave him a liberal education in languages, music, history, rhetoric, logic, mathematics and natural philosophy and also training in fencing and riding. But when the House of Orange lost its power, Constantijn lost a patron, and Christiaan lost his opportunity in diplomacy. Fortunately for us, Christiaan's true interests were in mathematics and natural philosophy. He developed the theory of pendulum motion and of colliding objects, invented an algorithm for computing the digits of the irrational number π, and constructed a telescope with which he identified Saturn's rings, its moon Titan, and the Orion nebula. But he is most remembered for his reflections on the nature of light.

The nature of light had long fascinated scientists. By the late 17th century two theories were current: (1) light was a stream of high-speed particles, and (2) light was a disturbance that propagated through an invisible medium called the ether. Descartes and

Newton, who agreed about little else, both supported the particle theory, while Huygens promoted the disturbance or wave theory.

Popular opinion was in favour of particles. After all, we can hear but not see around corners – facts that support the idea that sound is a wave disturbance that propagates through the air and around corners, while light is composed of small particles that, except at reflecting and refracting boundaries, travel in straight lines. But in 1660 the Jesuit Francesco Grimaldi (1618–1693) observed that light does, indeed, show a slight tendency to diffuse or *diffract* around small objects such as pins and through and around narrow slits in an opaque barrier, just as would be expected if light were a wave disturbance.

Huygens's *Treatise on Light* (1690) is quite narrow in scope. Huygens does not mention that white light is composed of a spectrum of colours, nor Grimaldi's discovery of diffraction, nor the properties we usually associate with waves such as periodicity and wavelength. Instead, Huygens asked: If light is a stream of particles, why don't the particles scatter from one another as different streams cross and, in this way, make ordinary vision impossible? And he observed that it is the nature of sound waves to propagate through each other without distortion. Otherwise, conversation in a noisy room would be impossible. Huygens concluded that light must be like sound: a disturbance or wave that propagates from source to receiver.

But a question remains. What is the medium through which light waves propagate? It cannot be air, since light travels through a glass jar from which the air has been pumped while sound does not. Evidently, light has its own medium that Huygens called the *ether*: an invisible material that penetrates transparent objects yet as a whole forms an elastic fluid through which disturbances propagate at high speed. The option of allowing light waves to propagate without a medium, the modern point of view, was not available to Huygens and other 17th-century scientists who were materialists and, as Huygens put it, dedicated to "... the true philosophy, in which one conceives the causes of all natural effects in terms of mechanical motions".

Huygens's contribution to the discussion begins by observing that every point on a luminous body is a source of light. And if the speed of light is the same in all directions, a disturbance originating from one point will, if unhindered, soon occupy the surface of a finite sphere centred on that point. But how exactly does one spherical disturbance evolve into another larger, concentric spherical disturbance? *Huygens's principle* according to which *every point in a disturbed medium is a new point source* provides the answer. Four points are identified on the inner circle of the left-hand diagram of figure 47 and each one of these, according to Huygens, sends out its own spherical disturbance, sometimes called a *secondary wave*. The envelope of these secondary waves (here identified with a dashed circle) constructs a new surface of disturbed medium. On the right-hand diagram, light propagates from a point source so distant that its spheres of disturbance appear in the diagram as parallel lines. Again, the envelope of secondary waves constructs a new line in the direction of forward propagation.

All the well-known properties of light – straight-line propagation in homogeneous media, equality of the angles of incidence and reflection, and Snell's law of refraction – follow from Huygens's principle. Consider, for instance, the wave surfaces depicted in figure 48 as they approach an air-water interface. Huygens supposed that the material through which light travels modifies the speed of light – the more dense the material, the slower the light. Therefore, the distance between the waves should shrink as they enter the water. One could imagine the secondary waves produced at the interface: two concentric half-circles, the one in water smaller than the one in air. But one could also, equivalently, imagine the wave crests as rows of a marching band closing up as the band leaves a smooth walking surface and enters a rough piece of ground – assuming that the march tempo remains steady while the marchers' pace shortens. The result: the lines of wave disturbance incline toward the normal of the interface in just the amount dictated by Snell's law given that the speed of light in water is about three quarters that in air.

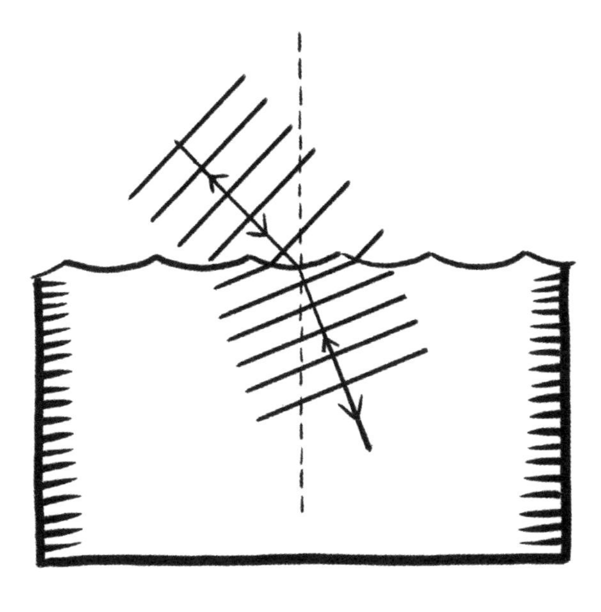

Figure 48

The particle and the wave interpretations of light coexisted throughout the 18th and early 19th centuries. In mid-19th century the speed of light in water was found to be three quarters that in air. Only Huygens's theory of light, which had by then evolved into a complete wave theory, was consistent with this result, even though direct evidence of the ether has never been found. Evidently light waves do not need a medium through which to propagate.

30. Bernoulli's Principle (1733)

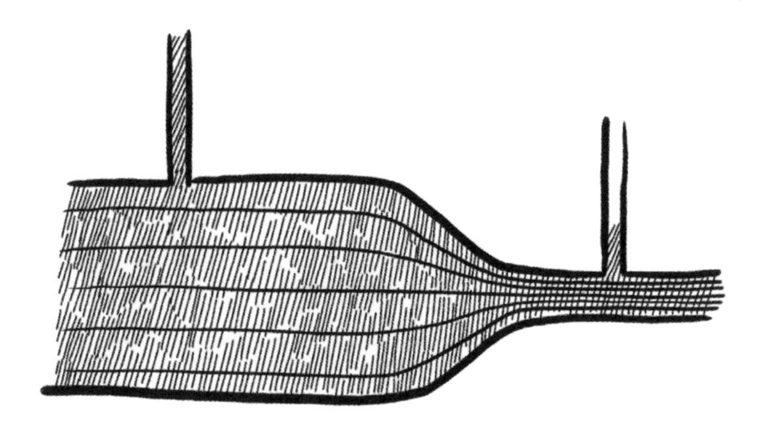

Figure 49

Daniel Bernoulli's (1700–1782) good fortune (and also his misfortune) was to have his own father as a tutor. Daniel's father, Johann Bernoulli (1667–1748), was a professor at the University of Basel in Switzerland and the foremost mathematician of Europe. Johann and his brother Jakob, Daniel's uncle, were among the first mathematicians to master the calculus after its invention by Newton and Leibniz in the second half of the 17th century. Johann's son, Daniel, was probably the ablest of several generations of Bernoulli mathematicians.

Such was Daniel's precocity and his broad talent that at 21 years of age he would have made the University of Basel a fine professor in any one of several fields: natural philosophy, mathematics, logic and physiology. But his application for a faculty position was passed over on two occasions, not because his qualifications fell short, but because the University chose the successful candidate by lot from among those qualified. Daniel was simply unlucky.

A decade later Bernoulli had forged an international reputation chiefly for his work at the Imperial Academy of Sciences in St Petersburg. He excelled at finding problems ripe for solution with the new methods. But he hated the harsh climate of St Petersburg and, in 1732, returned to Basel in order to accept the position finally offered him: a Professorship of Anatomy and Botany – two subjects in which, by that time, he had little interest. Only toward the end of his productive career did Bernoulli assume professorships that reflected his abiding interests: in physiology in 1743 and in natural philosophy in 1760.

Figure 49 illustrates one of Daniel Bernoulli's discoveries. The thick dark lines outline a section of tubing through which an incompressible fluid (for instance, water) flows from left to right. As the tube's diameter shrinks in the direction of fluid flow, the fluid speed increases in order that the fluid entering the tube from the left might leave on the right at the same rate. The thin lines are streamlines – reproducible trajectories of a lightweight object immersed in the fluid. Where the streamlines crowd together the fluid flows more quickly.

Leonardo da Vinci (1452–1519) had, much earlier, understood this behaviour. Evidently, Leonardo spent many pleasant afternoons dropping seeds into flowing brooks and watching their trajectories unfold in space and time – the seed speeding up and slowing down as the brook narrowed and widened. This simple inverse relation between a fluid's speed and the cross-sectional area of its channel, sometimes called the *principle of continuity*, expresses conservation of mass.

Daniel Bernoulli, who was well aware of the principle of continuity, searched for a second principle linking a fluid's speed to the pressure it exerts. Because Bernoulli was a former medical student he knew that measuring the pressure of a moving fluid, such as the pressure of arterial blood, presented a problem. Physicians in his day simply cut open a patient's artery and observed how high the blood spurted. Bernoulli sought a less wasteful and less dangerous method. He experimented with water flowing at various speeds through pipes of various diameters. He punched holes in the pipes and fitted these holes with vertical glass tubes open at both ends. When water flowed through the pipe, water ascended the tube. The pressure of the moving water equals the pressure of the ambient air plus an amount proportional to the height of the water supported in the open glass tube.

Bernoulli's technique quickly became standard medical practice. For the next 170 years physicians inserted the sharpened end of an open glass tube into a patient's artery and observed how high the blood ascended in the tube – the higher its ascent, the higher the patient's blood pressure. This method was better than the old one, but still painful and dangerous. Not until 1896 was the current non-invasive way of measuring blood pressure found.

More importantly, Bernoulli discovered that in each experimental arrangement the sum of the pressure exerted by the flowing fluid and the energy density of its bulk flow, $P+\rho V^2/2$ where P is the fluid pressure, V is its speed, and ρ is its mass per unit volume, remains constant along a streamline – a relationship we now call *Bernoulli's principle*. Evidently an incompressible fluid loses some of its pressure as it speeds up upon entering a narrower section of tube.

Today we use the principle of continuity and Bernoulli's principle to explain how aeroplane wings produce lift. Figure 50 shows the cross-section of an aeroplane wing and the streamlines of the air that flows around it. Straight, evenly spaced streamlines far above and immediately below the wing indicate undisturbed air. Immediately above the wing section the streamlines necessarily crowd together. According to the principle of continuity, the air immediately above the wing must flow more quickly than the air below it, while, according to Bernoulli's principle, the more quickly flowing air above the wing exerts less downward pressure on the wing than the more slowly moving air below the wing exerts upward pressure. The result is a net upward force on the wing.

Figure 50

Most fathers would be proud of a son with Daniel's accomplishments, but not Johann. Instead he saw Daniel as a competitor. When in 1734 the two tied for first place in a competition offered by the Paris Academy of Sciences to which they had submitted independent solutions to a problem in celestial mechanics, Johann angrily denounced his son – and also the prize committee for not recognizing his superior achievement. Then in 1743 Daniel discovered that his father Johann had reproduced or, as he suspected, plagiarized and published in *Hydraulica* (1743) much of what Daniel had published ten years earlier on moving fluids in his similarly titled *Hydrodynamica* (1733). Worse yet, Johann had asked the printer to backdate the publication of *Hydraulica* to 1732 in order to establish priority over his son. Daniel never forgave his father and the two remained unreconciled at the elder Bernoulli's death in 1748.

31. Electrostatics (1785)

Figure 51

Figure 51 illustrates a well-known interaction: *like charges repel and unlike charges attract*. The two kinds of charge are here indicated by a plus and by a minus sign. Insulating strings suspend the charged balls. (An *insulator* is made of material in which charges *are not* free to move, and a *conductor* is of material in which charges *are* free to move.)

In the 18th century these balls were composed of pith (a spongy organic material) and the strings were of silk – both good insulators. The natural philosophers of that time generated positive charge on a glass rod, for instance, by rubbing the rod with their hands, and transferring the charge to the pith balls by stroking the latter with the glass rod.

Today we can illustrate the same interactions more easily. Simply detach about 46 cm (18 inches) of Scotch tape from its spool. Fold one end of the tape over on itself to make a non-sticky handle and press the remaining sticky side to a smooth table surface. Prepare another tape in exactly the same way, and pull both tapes up from the surface at once – one in each hand. The identically prepared tapes are identically charged and will repel each other as shown in the right-hand diagram of figure 52. In order to create oppositely charged tapes, press one to the table and stick another on top of the first – sticky side to non-sticky side with the two handles at the same end. Pull the tapes up while stuck together. Then, carefully peel them apart. They will be oppositely charged and attract each other as shown in the left-hand diagram of figure 52.

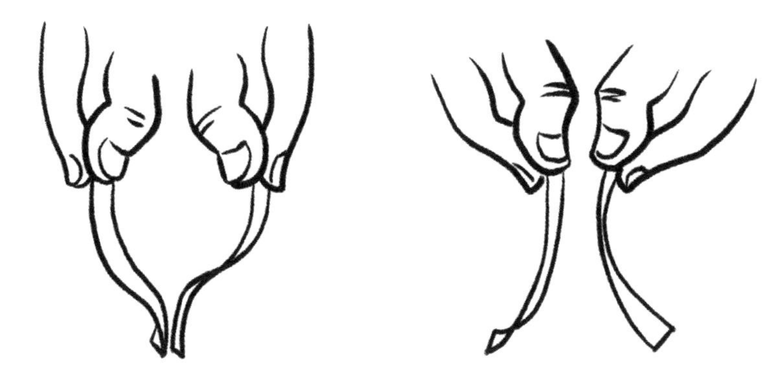

Figure 52

Any two dissimilar materials in close contact, such as tape and table surface, will result in charge moving from one material to the other. The traditional method of charging by rubbing is simply one way of making close, repeated contact between two dissimilar materials. Placing the sticky side of one tape on the non-sticky side of a second tape is another.

Benjamin Franklin (1706–1790) invented the names *positive* and *negative* to denominate the two kinds of charge, but they could have been given other names – and, indeed, were by Charles du Fay (1698–1739), who was the first to recognize the phenomenon of like charges repelling and unlike charges attracting. Du Fay called the charge created on glass by rubbing glass *vitreous* (after the Latin root for *glass*) and the charge created on amber by rubbing amber *resinous* (after the Latin for fossilized tree resin). In Franklin's jargon vitreous charge was *positive* and resinous charge *negative*.

The names *positive* and *negative* suggest Franklin's theory according to which all materials contain a single electric fluid. When an object has an excess of this fluid it is charged positive, and when an object has a deficit it is charged negative. Other people explained the same phenomena by appealing to two different kinds of fluid, each with its own charge. Compelling evidence in favour of two different charges (and their two different particles) did not emerge until the discovery of the electron in the late 19th and the proton in the early 20th century.

If you try the experiment with Scotch tape, you may notice that your hand attracts the tape regardless of its charge. The explanation is simple. Under normal conditions skin is a good conductor. As your hand approaches, for instance, a positively charged tape, negative charges within your hand move closer to the tape and positive charges move farther away. Your hand (actually your whole body) becomes *polarized*. Since the closer the charges the stronger their attraction, your hand attracts the tape independently of the kind of charge the tape contains. Figure 53 shows how, in similar fashion, an uncharged, horizontal conducting rod becomes polarized and attracts a vertical, positively charged tape.

Figure 53

The 18th century was the first age of electricity. It was also the age of enlightenment – a period when men and women believed that rational thinking, rather than adherence to tradition, would solve our problems and improve our lives. The natural philosophers of the 18th century not only discovered that like charges repel and unlike ones attract but invented means of mechanically generating large amounts of charge. Some went in for fantastic demonstrations that, for instance, polarized or charged small boys suspended from bundles of silk threads. Benjamin Franklin not only drew electricity from storm clouds with his wet (and so conducting) kite string, but also managed to cook a turkey by passing electric charges through it. Stephen Gray (1666–1736) conducted an electrostatic signal some 244 metres (800 feet) along a metal wire – a precursor to the telegraph. Others hawked the therapeutic effect of electric shocks.

Charles Coulomb's (1736–1806) contribution crowned the efforts of Gray, Du Fay, and Franklin. In order to accurately measure the strength of the Earth's magnetism Coulomb eliminated friction from a compass needle by suspending it from a fine thread. The more force exerted on the ends of the needle, the more it rotated and twisted the suspending thread. Coulomb used this technique to devise a sensitive device, now called a *torsion balance*, for measuring the force between two small, charged spheres. One sphere was attached to the end of a light, horizontal, counter-balanced rod suspended from a fine thread. The other stationary, charged sphere was placed near the first. The larger the force exerted on the charged sphere attached to the suspended rod, the more the rod rotated and turned its suspending thread – the rotation in direct proportion to the force applied. Coulomb found that the force F exerted between two small charges, q_1 and q_2, is inversely proportional to the square d^2 of the distance between them and directly proportional to the product of the charges – a result, $F \propto q_1 q_2 / d^2$, now known as *Coulomb's law*.

While this result, mimicking as it does Newton's universal law of gravitation, had been suggested before, Coulomb was the first to demonstrate it with a simple, convincing experiment. He did not speculate on the cause of this *electrostatic force*, but rather, like Newton before him, was satisfied with its precise mathematical description. Coulomb is one of the 72 notable French engineers, scientists and mathematicians whose names are inscribed on the Eiffel Tower.

Nineteenth Century

32. Young's Double Slit (1801)

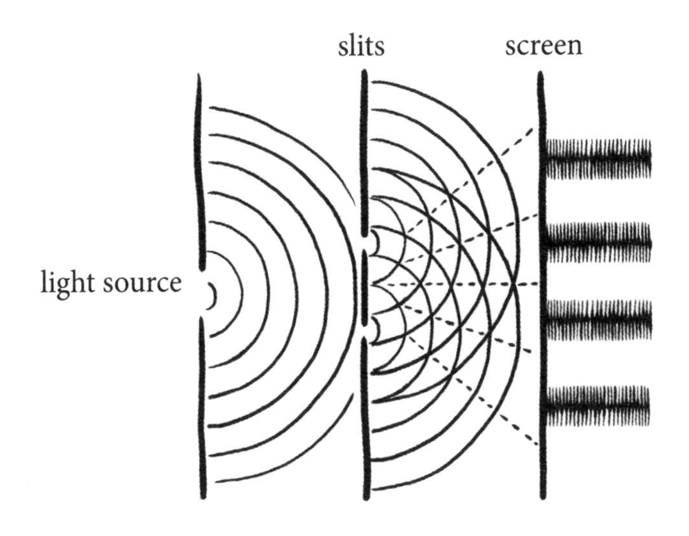

Figure 54

Sometimes one physical theory completely overthrows and replaces another. Such a revolution occurred in the years following the period 1801–04 when the English polymath Thomas Young (1773–1829) marshalled compelling arguments in favour of the wave theory of light. Young's arguments were, in part, reinterpretations of data gathered a hundred years earlier by Isaac Newton (1642–1727) and, in part, based upon his own simple experiments.

Young had to overcome a strong prejudice among natural philosophers in favour of Newton's hypothesis that light is composed of small particles that travel in straight lines at high speeds. True, Newton had, early in his career, explored the possibility that light was composed of waves. But the waves of which he was aware (sound and water waves) tend to bend, that is, to *diffract*, around barriers, while light, it seems, does not. After all, we can hear but not see around corners. Consequently, Newton was drawn to the idea of particles of light. In order to explain optical phenomena more complex than the formation of shadows, Newton endowed different kinds of light particle with different tendencies to transmit and reflect. These ideas were clearly speculative, but the unqualified success of Newtonian mechanics and gravitation gave them an unearned authority.

Young broke this century-long Newtonian spell by first noting that the intensity of sound and water waves, in fact, diminishes behind the barriers around which they diffract – just as does, in greater degree, the intensity of light. Evidently, all three phenomena (light, sound, and water waves) diffract around barriers in much the same way if in different degrees. Figure 54 illustrates the geometry of one of Young's demonstrations of diffraction – the *double slit geometry*. Young discussed several realizations of this geometry: one with water waves, one with light of a single colour, and one with white light composed of many colours.

It is easiest to see the physics behind all three realizations when the waves are on the surface of a pool of water. Young used what today would be called a "ripple tank", essentially a flat basin filled with water, to demonstrate how waves propagate around barriers and interact with one another. Waves are launched in the ripple tank by the periodic insertion and removal of a solid object. The dark lines in figure 54 indicate the crests of these waves. The wave troughs are, of course, midway between the crests. By design, the crest of a single wave strikes the two openings in the middle barrier at the same time and launches two new waves into the area to the right. (Recall Huygens's principle that each point on a wave front is a new point source.) Where, in the middle region, two dark lines intersect, two wave crests meet, superpose, and form a double-high crest. Where two troughs meet, a double-deep trough forms. And where a crest meets a trough, the water remains at its undisturbed level. Young coined the word *interference* to denote this pattern of wave interaction.

That light produces, on a much smaller scale, the same interference pattern in the double-slit geometry as do water waves is evidence that light is also composed of waves. In particular, monochromatic light originating from a single point source and propagating or diffracting through two parallel openings or slits in an otherwise opaque barrier produces a series of bright and dark bands, an interference pattern. Light, in effect, propagates along the direction indicated by the dashed lines of figure 54.

Young went on to describe the different interference patterns produced when light is diffracted around a fine thread and reflected from grooved surfaces and thin films. Young concludes one of his (typically wordy) lectures with the statement that

> . . . the accuracy, with which the general law of interference of light has been shown to be applicable to so great a variety of facts, in circumstances the most dissimilar, will be allowed to establish its validity in the most satisfactory manner.

Evidently, since interference is a wave property and light can be made to produce various interference patterns, light is composed of waves.

Because Young's printed arguments were entirely verbal – that is, without benefit of diagrams or mathematics – they were at first ignored. Eventually, Augustin Fresnel (1788–1827) provided the mathematical formulation implied by but lacking in Young's presentations. Sometimes Young's arguments have been interpreted as "proving Newton wrong". If so, Young also honoured Newton for his enduring contributions.

Young was a child prodigy, the eldest of ten children born to Quaker parents of modest means. He was raised by his grandfather and educated by an aunt who, for the most part, allowed him to pursue his own interests. He read with fluency by the age of two and at four had twice read through the entire English Bible. Besides

European and classical languages he studied Near Eastern ones: Hebrew, Samaritan, Chaldean, Syriac and Persian. He kept a journal in Latin and commented on French authors in French and on Italian authors in Italian. Once when asked to exhibit his penmanship he wrote the same sentence in fourteen different languages. While Young trained to be a physician, he also developed a serious interest in mathematics; natural philosophy, in particular optics and botany; and various mechanical arts including telescope making.

Young's competence in languages led him to study a copy of the three inscriptions – one in Ancient Greek, one in Egyptian hieroglyphics, and one in demotic Egyptian – on a particular stele, the so-called Rosetta Stone, discovered in 1799 by a French officer with Napoleon's army in Egypt. Since all three inscriptions paraphrase the same decree, the Rosetta stone was a key to deciphering Egyptian hieroglyphics – the meaning of which had been, since the late Roman period, lost. Young's contribution was to discern that the demotic was a mixture of alphabetic and hieroglyphic characters and to begin the work of deciphering both Egyptian texts. When, in 1822, the French philologist Jean-François Champollion (1790–1832) independently deciphered the demotic and hieroglyphic inscriptions, Young praised his work.

Young's life of scholarship earned him his own stele, a memorial stone in Westminster Abbey – one that praises

> a man alike eminent in almost every department of human learning, patient of unremitting labour, endowed with the faculty of intuitive perception, who, bringing an equal mastery to the most abstruse investigations of letters and science, first established the undulatory theory of light, and first penetrated the obscurity which had veiled for ages the hieroglyphics of Egypt.

33. Oersted's Demonstration (1820)

Figure 55

Once household implements began to be made of iron, people began to notice that nearby lightning strikes sometimes magnetized these implements. But what is lightning? And how does it magnetize iron? In letter written in 1752 Benjamin Franklin (1706–1790) described an experiment whose purpose was to answer the first of these questions. His idea was to fly a kite into a storm cloud so that any electrical charge present would be conducted down its wet string and stored in a glass container, lined inside and out with metal foil, called a *Leyden jar*. After explaining to his correspondent how to make a kite out of a silk handkerchief, Franklin went on to say,

> And when the rain has wet the kite and twine, so that it can conduct the electric fire freely, you will find it stream out plentifully from the key on the approach of your knuckle. At this key the phial [Leyden jar] may be charged, and from electric fire thus obtained, spirits may be kindled, and all the other electric

> experiments be performed which are usually done by the help of a rubbed glass or tube, and thereby the sameness of the electric matter with that of lightning completely demonstrated.

One can only conclude, as Franklin does, that lightning consists of moving electric charges.

It was left to the Danish scientist Hans Christian Oersted (1777–1851) to address the second question: the relationship between moving charge and magnetism. The story goes that in 1820, while attempting to show his students at the University of Copenhagen that a current of moving charges had nothing to do with magnetism, Oersted placed a segment of conducting wire close to and parallel with the usual north-south orientation of a compass needle. To his surprise, when the ends of the wire were connected to a Voltaic cell or battery, the needle rotated away from the north-south line – as shown in figure 55. In actual fact, Oersted had unintentionally demonstrated that moving charges do produce a magnetic influence. The entire scene – Oersted, current-carrying wire, compass needle, and attentive students (all male) – is represented on one side of the Oersted Medal, awarded annually by the American Association of Physics Teachers to an outstanding teacher of physics. Oersted's image is certainly appropriate for this medal since his may be the only major scientific discovery made during a lecture demonstration before a class of students.

What most interested Oersted and his contemporaries was that a current of charge in the wire *twists* the compass needle. And, in this experimental arrangement, the twisting could not be explained in terms of an attraction or repulsion along lines joining points on the conducting wire and points on the compass needle. The only other known fundamental forces, Newton's law of universal gravitation and Coulomb's law of electrostatics, act in this way: that is, along lines connecting two points.

Oersted found that if the current in the wire was moderately strong, the magnetic influence it produced overcame that of the Earth. In this case, a set of compass needles arranged in a plane perpendicular to the wire would point in directions that circle the wire, as shown in figure 56. According to Oersted,

> From the preceding facts we may likewise collect that this conflict [or influence] performs circles; for without this condition it seems impossible that the one part of the uniting wire, when placed below the magnetic pole, should drive it towards the east, and when placed above it towards the west; for it is the nature of a circle that the motions in opposite parts should have an opposite direction.

What "performs circles" around the wire was later identified as a *magnetic field line*. Oersted had, in effect, demonstrated that an electric current produces a magnetic field and, in so doing, initiated the study of *electromagnetism*.

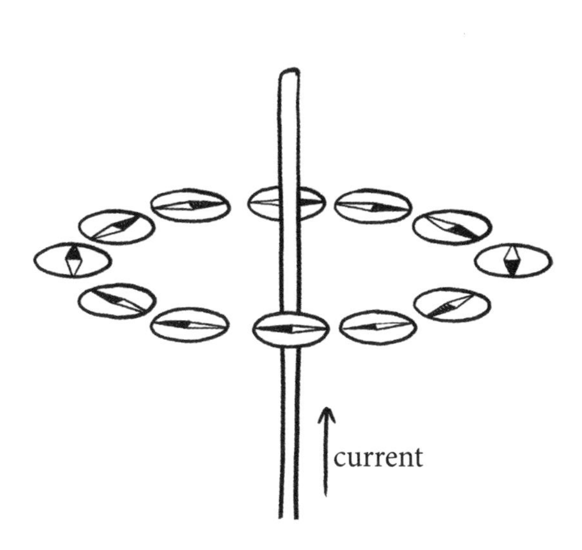

Figure 56

The connection Oersted revealed between electrical and magnetic phenomena satisfied the expectations of the contemporary Romantic Movement – a movement that touched on all aspects of natural phenomena and human endeavour. A romantic would see connections everywhere, and a scientifically inclined romantic would imagine that all the forces of nature were but different aspects of a single, all-encompassing, invisible power. If this power encompassed the human as well as the natural, then the natural world might suggest to us what it means to be human, and our humanity might teach us how to appreciate and care for the natural world. Oersted would have embraced these possibilities – so beautifully expressed in William Wordsworth's sonnet *The Prelude* (1850).

> My heart leaps up when I behold
> A rainbow in the sky:
> So was it when my life began,
> So is it now I am a man,
> So be it when I shall grow old
> Or let me die!
> The child is father of the man:
> And I could wish my days to be
> Bound each to each by natural piety.

But a romantic awareness could terrify as well as inspire. One need only read Mary Shelley's *Frankenstein* (1818) or Robert Louis Stevenson's *The Strange Case of Dr Jekyl and Mr Hyde* (1886).

Oersted was a child of his age, a romantic, and of the last generation of scientists who called themselves *natural philosophers*. In addition to his work as a physicist and chemist (he was the first to isolate the element aluminium) Oersted wrote a dissertation on Kantian metaphysics and published a volume of poetry. His last work was a philosophy of life entitled "The Soul in Nature".

34. Carnot's Simplest Heat Engine (1824)

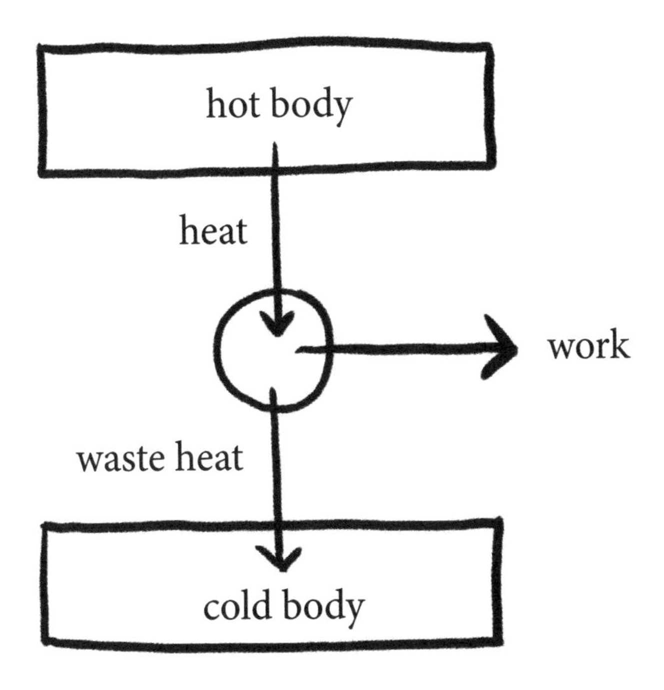

Figure 57

Steam-driven heat engines turned wheels that, in the early 19th century, ground corn, wove cloth, moved goods, and lifted water out of English coal mines. By the late 19th century heat engines were powering dynamos that produced electricity – that highly transportable potential to perform work. Remarkably, already in 1824 with the publication of *Reflections on the Motive Power of Fire and on the Machines Fitted to Develop that Power*, Sadi Carnot (1796–1832) had outlined the general possibilities and absolute limitations of heat engines.

Carnot's *Reflections* is concerned not only with the theory of heat engines and with their various applications, but also with the military, political, and economic implications of their development. This is not surprising given Carnot's family background. His father, Lazare, was Napoleon's capable general-in-chief, and Sadi's early military and scientific training was at the newly established Ecole Polytechnique. Fatefully for Carnot, England, rather than his native France, had discovered, developed, and applied the steam engine to the point that

> To take away today from England her steam-engines would be to take away at the same time her coal and iron. It would be to dry up all her sources of wealth, to ruin all on which her prosperity depends, to annihilate that colossal power. The destruction of her navy, which she considers her strongest defence, would perhaps be less fatal.

By 1824 that "colossal power" had exiled Sadi's father and blighted his own military career. France was worthy of a better future and a bigger role in developing the heat engine – or so Carnot must have thought. As it happened Carnot helped create that role by developing the theory of heat engines with a precision never imagined by the English engineers of his time.

The block diagram shown in figure 57 particularly suits the generality of Carnot's theory. Gone are the furnaces, boilers, pistons, condensers, and smoke stacks that compose a real steam engine. In his imagination Carnot stripped all these away until left with only the three elements and their functions that were essential for the operation of any imaginable heat engine: a hot body that supplies heat, a device that produces work from that heat, and a cold body that absorbs waste heat. Two blocks, one circle, and three arrows represent the elements and functions that compose Carnot's simplest heat engine.

That both a hot and a cold body are needed for a heat engine to produce work is the crucial contribution of Carnot's *Reflections.* He must have been aware of the importance of this requirement because he repeated it seven times in seven successive paragraphs within the first few pages of the *Reflections.* In Carnot's words

> The production of motion in steam engines is always accompanied by a circumstance on which we should fix our attention. This circumstance is the re-establishing of equilibrium in the caloric; that is, its passage from a body in which the temperature is more or less elevated, to another in which it is lower.

Eliminate either the hot or the cold body and whatever is left is no longer a heat engine capable of doing work. The logically equivalent statement that *No heat engine simpler than Carnot's simplest heat engine is possible* is a truth of the highest order – a truth that expresses what we now call the second law of thermodynamics.

Other versions of the second law of thermodynamics are better known. For example, *No process is possible whose only result is to cool a cold body and heat a hot body.* In other words, a cup of hot coffee left in a cool room never gets hotter; it always cools down. And, *No process is possible whose only result is to cool a hot body and produce work.* In other words, a heat engine cannot be 100% efficient. The German physicist Rudolph Clausius framed the first of these statements in 1850, and the English physicist William Thomson the second in 1851. Of course, Carnot's earlier 1824 statement pre-dates both. But each is logically equivalent to the other two. Each is a version of the second law.

One does not prove a statement that purports to be as fundamental as the second law of thermodynamics. Rather, one simply asserts that statement and from it derives important consequences. Only after these consequences have been tested and verified is the statement recognized as a fundamental law of physics.

Carnot was, indeed, able to derive important consequences from his version of the second law. For instance, he found that, in principle, *the most efficient heat engine is one that operates indefinitely slowly, without friction or dissipation, and without direct contact of hot and cold parts*. The technical word that stands for this combination of properties is *reversibility*. Thus, Carnot proved that *the most efficient heat engine is one that operates reversibly* – a statement traditionally called "Carnot's theorem".

Carnot was hampered in developing the consequences of his ideas because, at the time of writing the *Reflections*, he did not accept the law of conservation of energy, now known as the first law of thermodynamics. (Interestingly, the second law pre-dates the first law by more than 20 years.) In its place Carnot believed that heat, or *caloric* as it was called, was an indestructible fluid whose quantity was conserved as it flowed from one place to another. Not until the 1840s did James Joule's increasingly precise experiments explode the concept of caloric and compel acceptance of the first law of thermodynamics. According to the first law, it is energy, rather than caloric, that is conserved. In this view heat is just one way of transferring energy from one place to another. (Doing work is another.) However, such were Carnot's gifts that, even under the fog of serious misconception, he recognized and exploited a truth of great consequence – a truth we now call the second law of thermodynamics.

35. Joule's Apparatus (1847)

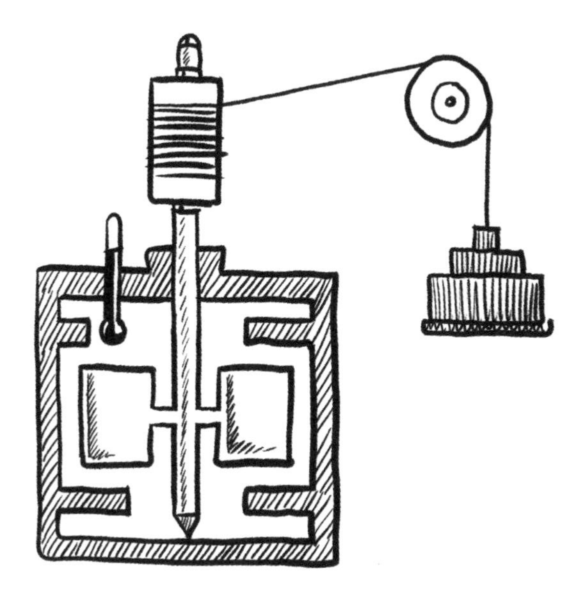

Figure 58

What causes things to heat up and to cool down? Thanks to the widespread use of reliable thermometers in the 18th century, one scientist came up with an explanation. According to Antoine Lavoisier (1743–1794), heating was caused by the flow of *caloric* (a "subtle fluid") that, as it penetrated the pores of an object, raised its temperature. Caloric was ingenerate and indestructible – that is, conserved – as it flowed from one object to another. Furthermore, caloric was weightless and composed of particles that repelled one other. Hot objects were caloric-rich and cold ones caloric-poor. As caloric diffused from a hot object to a cold one their two temperatures approached one another.

A particular amount of a particular kind of material requires a particular quantity of caloric to raise its temperatures by one degree. In this way, different objects have different *heat capacities*. The common substance water provides a convenient standard of comparison. By definition, *one calorie* is that quantity of caloric required to raise the temperature of one gram of water one degree Celsius (or Centigrade). Therefore, the heat capacity of water is, by definition, one calorie per gram degree Celsius.

If caloric merely flows from one place to another, one might, with the help of a table of heat capacities, predict temperature changes in a whole class of phenomena. Pour some cold milk into hot coffee. Given the amounts of milk and coffee, their heat capacities (essentially that of water), and their initial temperatures, the final temperature of the mixture follows from the conservation of caloric. Simple calorimetric experiments done in elementary physics and chemistry labs use the same principle.

However, the concept of caloric was far from universally accepted in 1800. Benjamin Thompson's cannon boring experiment of 1798 had seriously undermined, without altogether discrediting, the concept of a conserved caloric. Thompson (1753–1814), later known as Count Rumford, was an American original, sharp but self-aggrandizing. Born in Woburn, Massachusetts, his sympathies shifted to the British during the Revolutionary War, and when the tide turned in favour of the new nation Thompson left the rich widow he had married and resettled in England. Within a few years King George III had knighted Thompson, and, with the King's blessing, he became a scientific and military advisor to the Elector of Bavaria, all the while continuing to spy for his British patrons.

It was in this position that, when supervising the boring of cannon in Munich, Thompson, now Count Rumford of the Holy Roman Empire, reflected upon the necessity of continually cooling the cannon with water. To further investigate he devised an experiment in which a blunt tool bored into a cylinder of brass while the whole was immersed in the water contained within a sealed wooden box. As two horses turned the bore he noticed a continual increase in the temperature of the water until, after two and a half hours, it began to boil. As Rumford noted, in a report to the Royal Society of London (1798),

> It would be difficult to describe the surprise and astonishment expressed in the countenance of the bystanders, on seeing so large a quantity of cold water heated and actually made to boil without fire.

From whence came all this caloric? That caloric was released from the metal when shaved from the stock seemed unlikely since the heat capacity per mass of the metallic shavings was identical to that of the stock. Whatever its source, the supply of caloric seemed inexhaustible. According to Rumford,

> It is hardly necessary to add that anything which any insulated body, ... can continue to furnish without limitation cannot possibly be a material substance: and it appears to me to be extremely difficult, if not quite impossible, to form any distinct idea of anything capable of being excited and communicated, in the manner the heat was excited and communicated in these experiments, except it be MOTION.

But Rumford's view was not compelling. For his idea that heat is motion and stored in the motion of the smallest parts that compose a material was not easily quantified. Then again, the concept of caloric led to numerical predictions that worked – at least in calorimetric experiments. Undermining a theory is one thing. Replacing it with a sufficiently explanatory alternative is another.

It was not until 1847 that the young English brewer and amateur scientist James Joule (1818–1889) perfected a demonstration that demolished the concept of caloric, demoted calorimetry to a special case, and established the more broadly defined quantity *energy* as that which is always conserved. Since 1839 Joule had been labouring to show that given amounts of work generate determinate amounts of caloric – work that was variously performed by electrical currents, by rubbing surfaces, and by compressing gases. In each case Joule found that the same quantity of work, of whatever kind, generates the same amount of caloric. In the English units of his day, the energy required to lift approximately 780 pounds one foot generated the caloric necessary to raise the temperature of one pound of water one degree Fahrenheit. But if caloric could be generated at fixed, determinate rates, then caloric was not a conserved quantity.

Joule exhibited the apparatus depicted in figure 58 at a meeting of the British Association for the Advancement of Science in Oxford in 1847. It consisted of a well-insulated container of water into which a paddle wheel, driven by a falling weight, is inserted. Stationary vanes attached to the container keep the water from persistent circulatory motion. In this way the potential energy of the weight is dissipated in the water – the effect of which is soon indicated by a thermometer. Joule allowed the weight to drop again and again until the temperature rose a fraction of a degree – a precision with which, as a professional brewer, he was well acquainted – and claimed that the result was reproducible. Joule's simple design and precise measurements earned what Rumford's experiment had not: the attention of the English scientific elite. Indeed, one of those attending the 1847 meeting was William Thomson, who in 1851 adopted conservation of energy as a postulate of the new science of thermodynamics.

Joule continued to refine his measurement of the ratio of the work to the heat it produced, now called the *mechanical equivalent of heat*. His last measurement of this ratio (in 1878) yielded the number *772.55*, which is inscribed on his tombstone along with a verse from the gospel of John (9:4), "I must work the works of him that sent me, while it is day: the night cometh, when no man can work."

36. Faraday's Lines of Force (1852)

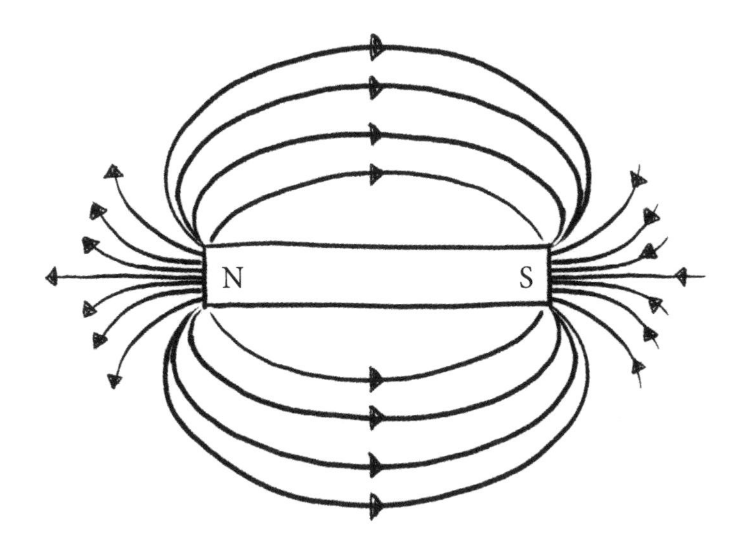

Figure 59

One of Albert Einstein's earliest memories was of a compass his father had given him. Apparently, the Earth, itself a large magnet, cast its influence across empty space and caused the compass needle to point north. Years later Einstein said that on holding the compass he "trembled and grew cold. . . . There had to be something behind objects that lay deeply hidden".

But is the space around a magnet really empty? Michael Faraday (1791–1867) was the first to gather evidence suggesting that what surrounds a magnet is as real as the magnet itself. He referred to this something as the "atmosphere" of a magnet or, alternatively, as its "lines of force". Today we speak of its *magnetic field*.

Shortly after Hans Christian Oersted discovered, in 1820, that a wire carrying a current deflects a magnetized needle, Faraday, a self-taught English chemist and physicist, began a three-decade-long study of electromagnetic phenomena. He reported the results of his study in 3,299 consecutively numbered paragraphs that occupy some 1,100 pages of text collected in three volumes called *Experimental Researches in Electricity*. His final contribution to this extraordinary work is an essay, "The Physical Character of the Lines of Magnetic Force" (1852), in which he expressed his belief in the physical reality of lines of magnetic force.

It is not difficult to map a bar magnet's lines of force with a small compass. One has only to put the magnet in the centre of a large sheet of paper and position the compass nearby. Place a dot at the compass needle's head (or north pole), shift the needle's tail (or south pole) to the position of the dot, and place another dot at the new position of the needle's head. Repeat this process many times and smoothly connect the dots. By convention, magnetic lines of force have a direction. They begin at a north pole and end at a south pole. The pattern of lines drawn in this way and according to this convention will approximate the idealized pattern shown in figure 59. Knowing the direction of the lines of force surrounding a magnet is equivalent to knowing in which direction a compass needle will point when placed near the magnet.

According to Faraday, magnetic lines of force have the following properties: (1) lines of force tend to shorten themselves, (2) adjacent parallel lines of force pointing in the same direction repel each other, and (3) adjacent parallel lines of force pointing in opposite directions attract each other and then reconnect or merge. The diagrams in figure 60 illustrate the arrangement and suggest the behaviour of the lines of force associated with particular magnets and current-carrying wires. In the leftmost diagram the lines of force shorten and cause north and south poles to attract each other. In the middle diagram adjacent parallel lines of force pointing in the same direction repel and cause two north poles to repel each other. Finally, the rightmost diagram shows the cross-sections of two wires both with electrical currents flowing out of the plane of the paper. Adjacent parallel lines of force pointing in opposite directions in the region between the two wires attract each other, merge, shorten, and cause the two wires to attract.

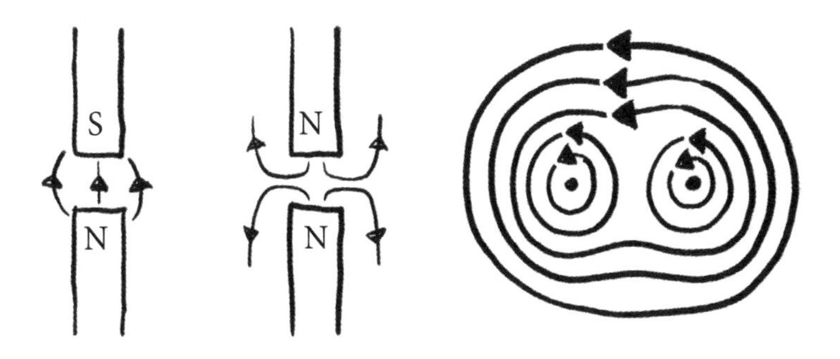

Figure 60

Faraday knew that his lines of force explained nothing that could not also be explained in terms of distant objects exerting forces across empty space. The lines of force are, as he admitted, "speculations" rather than deductions along a "strict line of reasoning". Even so his lines of force are immensely satisfying. Because adjacent lines of force, or adjacent parts of the same lines, push or pull directly on each other, the lines eliminate the need for action-at-a-distance forces. Lines of force helped Faraday, and also help us, visualize what happens. While Faraday was perhaps the most productive experimental physicist of all time, his mathematical knowledge did not extend beyond elementary algebra and trigonometry. Visualizations, of which the magnetic lines of force are a prime example, did the work of mathematics for Faraday.

Faraday was born to a family so poor, he often went hungry as a child. He was raised in and remained devoted to a small, dissenting (that is, non-Anglican) Christian denomination. Although he had no schooling, he was apprenticed to a kindly bookbinder who encouraged Faraday in his education. The young Faraday attended public lectures on topics in natural philosophy. These helped him, at age 21 in 1813, to land a menial job at the Royal Institution where eventually he was able to do his own experiments.

His invention of the electric motor (1821) and discovery of electromagnetic induction (1831–1832) brought him recognition, but his relations with the scientific establishment of his day were complicated. Although he was universally honoured for his inventions, discoveries, popular lectures, and public service and he corresponded with the important scientists of his time, Faraday's theories and speculations were generally dismissed. Faraday had no pupils and no disciples apart from James Clerk Maxwell (1831–1879). He refused a knighthood and the presidency of the Royal Society, and declined to advise the British government on creating chemical weapons for use in the Crimean War (1853–1856). He died in 1867 before his lines of force and the concept of an electromagnetic field to which they gave birth were widely accepted.

Maxwell vindicated Faraday's work by translating Faraday's lines of force into mathematical language and incorporating that mathematics into a set of equations, Maxwell's equations, that compose a complete theory of electromagnetism. According to this theory electric and magnetic lines of force, while produced by charges, magnets and electrical currents, may detach from these sources and propagate with finite speed through empty space as electromagnetic waves: for instance, from the Sun to the Earth and from satellite to cell phone.

37. Maxwell's Electromagnetic Waves (1865)

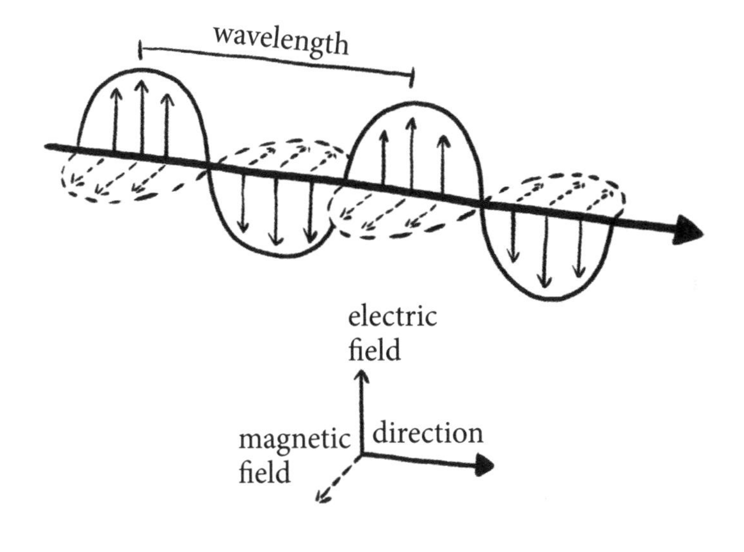

Figure 61

James Clerk Maxwell (1831–1879) so greatly admired Michael Faraday (1791–1867) he advised readers of his own work first to carefully study Faraday's 1,100-page *Experimental Researches in Electricity* (1855). Certainly he had done so and even corresponded with its author, 40 years his senior. Eventually, Maxwell paid Faraday the high compliment of constructing a mathematical model of the pictorial concept of which Faraday was most proud: his electric and magnetic lines of force.

Four equations, the celebrated *Maxwell's equations*, encapsulate Maxwell's model of Faraday's lines of force. These equations show how electric and magnetic lines of force, or *fields*, are generated from their sources: that is, from charges and currents. In promoting

Faraday's lines of force, Maxwell, like Faraday before him, was at odds with the many physicists committed to the programme of action-at-a-distance forces: that is, committed to explaining electromagnetic phenomena in terms of charges, in various states of rest and motion, exerting forces on other charges. Maxwell pointedly ignored forces and made fields his priority.

But a question remained. Are the lines of force and the fields to which they are equivalent real, as Maxwell and Faraday believed, or are they mere mathematical devices whose only purpose is to make convenient the calculation of forces? For a while the question remained unanswered. But in the process of constructing his equations Maxwell discovered something Faraday had missed. Not only do magnetic fields that change their intensity, direction or position generate electric fields as expressed by Faraday's law (one of the four Maxwell equations), but also electric fields that change their intensity, direction or position generate magnetic fields, an effect Maxwell incorporated into the Maxwell-Ampere law (another of the four Maxwell equations).

Together these two effects make possible self-sustaining electromagnetic waves that propagate at a speed determined by a combination of constants inherent in electromagnetic phenomena. Maxwell noticed that their predicted speed is close to measured values of the speed of light (metres/second). Furthermore, electromagnetic waves carry energy and momentum just as do light waves. Maxwell reasonably concluded that light is composed of electromagnetic waves. In this way, Maxwell established, in 1865, a new explanation of light and endowed electromagnetic fields with physical reality. Then in 1886–87 Heinrich Hertz experimentally discovered that electromagnetic waves behave in the same way as does light and, in this way, confirmed Maxwell's conclusion.

Figure 61 illustrates an electromagnetic wave, composed of electric and magnetic fields, each sustaining the other, and propagating in the direction shown by the thick black arrow. The whole structure looks three-dimensional, but this is illusory for only one dimension of space is shown. At every point along this single spatial dimension the electric and magnetic fields composing the wave have amplitudes, indicated by the length of their arrows, and a direction, indicated by the thick arrow's direction. The particular waveform shown also has a definite wavelength. In general, more complex waveforms are sums of a number of waves each with its own amplitude, direction and wavelength.

However revolutionary his discovery, Maxwell adopted the common sense of his time in supposing that electromagnetic waves require a material medium through which to propagate. After all, the waves known to Maxwell (sound waves, water waves, and waves in and on musical instruments) were waves that propagate in material media. Therefore it was natural for Maxwell and others to assume that electromagnetic waves also have a material medium – then called the *luminiferous ether*. Even so, this ethereal medium avoided and continues to avoid detection. Furthermore, its properties are incoherent. The ether must be very tenuous because the planets move through it without noticeable resistance. Yet the ether must also be rigidly elastic, like steel, otherwise the speed of light would not be so high. Rather than continue to believe in a material like no other whose only purpose is to be a medium for electromagnetic waves, physicists, toward the end of the 19th century, simply abandoned the ether.

Maxwell, like many from 19th-century, middle class, English families, received his early education at home under the guidance of parents and private tutors. As a child all things that moved, made a noise, or in some way "worked" fascinated Maxwell and prompted the question "What's the go o' it?" or the urgent request "Show me how it doos." He learned to draw from his cousin, Jemima Blackburn, who later became a well-known watercolourist and book illustrator.

Maxwell wrote scientific and mathematical papers before he was old enough to read them, as was then the custom, before the learned societies that published them. As a young scholar he won the Adams Prize (in 1857) for an extended analysis of the stability of Saturn's rings. Besides his contributions to electromagnetic theory, Maxwell developed a mathematical description of the range of particle velocities in a gas in equilibrium at a given temperature – the so-called *Maxwell distribution* – and made lasting contributions to colour vision, thermodynamics and statistical mechanics.

Maxwell died of abdominal cancer at the age of 48. Maxwell's admiring friends and colleagues mourned his early death. One of them, Professor Lewis Campbell, a friend from childhood, wrote a biography (1882) that emphasized Maxwell's Christian faith. Maxwell, Campbell said, had taken to heart his mother's request that he "look up through Nature to Nature's God".

Twentieth Century and Beyond

38. Photoelectric Effect (1905)

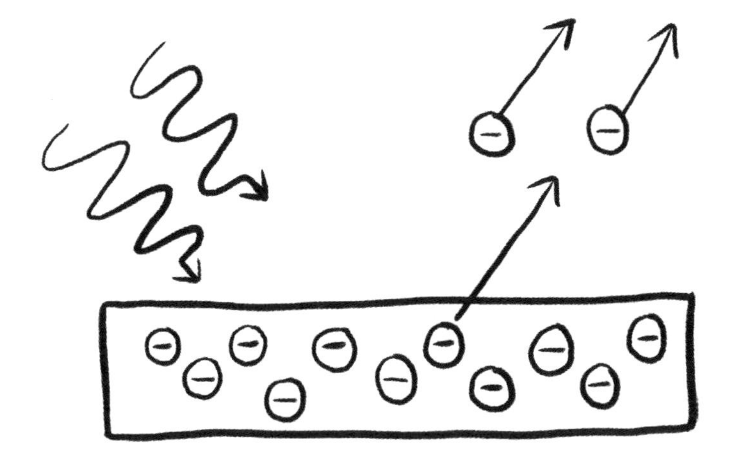

Figure 62

Figure 62 depicts the *photoelectric effect*. Light (at the upper left) strikes a surface (at bottom), breaks loose some of its electrons, and ejects them (at the upper right) from the surface. Soon after its discovery in 1887 scientists began exploring the curious properties of the photoelectric effect. Chief among them is that only sufficiently high frequency light can eject electrons – so called *photoelectrons* – from the surface. How high a frequency is required depends on the composition of the surface. Most metals, for instance, require frequencies at least as high as that of ultraviolet light. If the frequency is too low – for instance, if its colour is too red – no photoelectrons are produced no matter how intense the light. Then, in 1902, Philipp Lenard (1862–1947) discovered that when photoelectrons are produced their kinetic energy increases with the frequency of light that produced them and is independent of its intensity.

This behaviour is impossible to explain in terms of the wave theory of light. All simple waves are characterized by a frequency, which determines how quickly wave crests arrive at a particular point, and by an amplitude, whose square is directly proportional to the wave intensity. Ocean waves, for instance, have these properties.

Indeed, Arthur Holly Compton (1892–1962) once made clear the absurdity of low-amplitude light waves ejecting electrons from a metal with the following simile.

> There was once a sailor on a vessel in New York harbor who dived overboard and splashed into the water. The resulting wave, after finding its intricate way out of the harbor, at last found its way across the ocean, and a part of it entered the harbor at Liverpool. In this harbor there happened to be a second sailor swimming beside his ship. When the wave reached him, he was surprised to find himself knocked by the wave up to the deck.

Here Compton actually makes two comparisons. The first two sentences refer to the creation of electromagnetic radiation by electrons striking a metallic surface (see essay 41), while the last two sentences refer to a single electron absorbing, photon-like, low-amplitude electromagnetic radiation.

In 1905 Albert Einstein (1879–1955), then a 26-year-old Swiss patent inspector, devised a simple explanation of the photoelectric effect. According to Einstein, light has both a wave-like and a particle-like character. Light must be wave-like because Maxwell's wave theory of light had been tremendously successful. Yet, Einstein argued, in producing photoelectrons, the energy of light behaves as if it is concentrated in bundles or *quanta* (later called *photons*). The energy of a photon $h\nu$ is proportional to the frequency ν of the wave with which it is associated. The proportionality constant h is called *Planck's constant* after the physicist Max Planck (1858–1947), who first measured its value. Greater light intensity simply means greater numbers of photons.

Since a photon is localized, one photon interacts with only one electron. Part of the photon energy goes into overcoming the forces that hold the electron in the surface, while the other part produces the kinetic energy of the electron. Symbolically $hv=W+E_k$. Thus, if the energy of the photon is smaller than the energy required to dislodge an electron, that is, if $hv<W$, no photoelectrons are created. The data available to Einstein supported his light quantum or photon interpretation of the photoelectric effect.

In 1900 Max Planck derived an accurate description of *equilibrium* or *blackbody radiation*: that is, a description of the electromagnetic waves contained within a cavity whose walls absorb and emit those waves. In doing so he supposed that the material composing the cavity walls absorbs and emits electromagnetic wave energy only in quantized chunks of energy hv. Einstein's explanation of the photoelectric effect shifted attention away from the way blackbody radiation interacts with the material composing the cavity walls to the radiation itself.

The title of Einstein's 1905 paper on the photoelectric effect, "On a Heuristic Point of View about the Creation and Conversion of Light", signals Einstein's cautious approach. Photons are a mere *heuristic*: that is, a useful, but ultimately provisional, approach to the photoelectric effect. Einstein came to understand that photons and other quantum phenomena presume a probabilistic dynamics. (The creation of a photoelectron is not completely predictable.) But Einstein never accepted as final the probabilistic interpretation of physics that the quantum revolution demanded.

Planck was not pleased with photons for a different reason. He asserted that if photons were accepted "the theory of light would be thrown back by centuries" – presumably back to the 17th century when the adherents of light particles (following Newton) and of light waves (following Huygens) debated the issue. Thus, Planck understandably resisted the wave-particle duality of light. But wave-particle duality was here to stay. Sometimes it is asserted that the current theory of light, called *quantum electrodynamics*, as developed after World War II, decides the question in favour of particles. If so, these are very strange particles that carry along with them information usually ascribed to waves.

Robert Millikan (1868–1953) performed experiments in 1915–16 that confirmed Einstein's explanation of the photoelectric effect. Einstein received the 1921 Nobel Prize (in 1922) "especially for his discovery of the law of the photoelectric effect", and Millikan received the 1923 Nobel Prize, in part, for his work in confirming that law. Einstein, however, was never satisfied with the photon concept. In 1951 he wrote that, "All these 50 years of pondering have not brought me any closer to answering the question, 'What are light quanta?'"

39. Brownian Motion (1905)

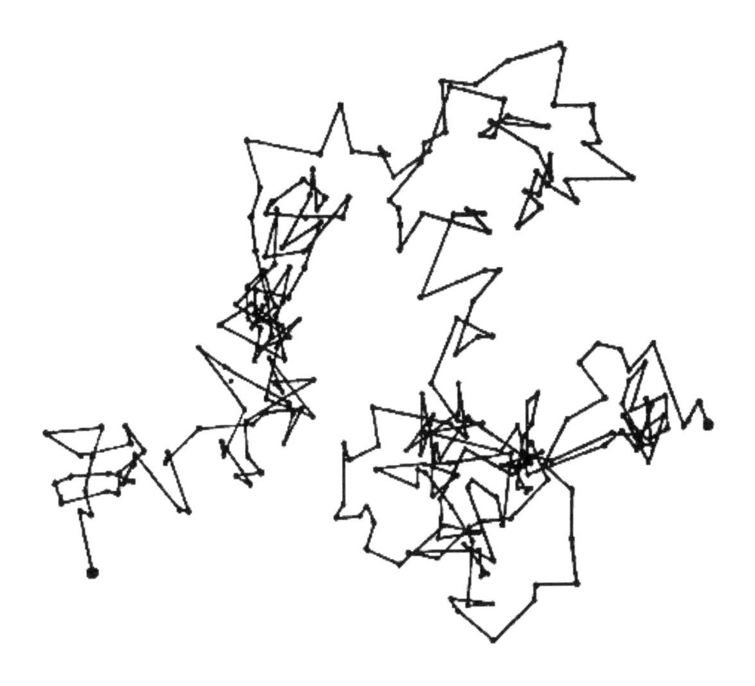

Figure 63

Jean Perrin's (1870–1942) experimental work ended the long debate over whether matter was continuously divisible or not: that is, whether or not atoms exist. Perrin received the 1926 Nobel Prize in physics for deciding the question in favour of atoms. Among his crucial experiments are those that confirmed Albert Einstein's theory of Brownian motion – a theory that makes use of atoms and molecules.

Brownian motion – that irregular, back and forth, wandering motion of microscopic particles immersed in a liquid – was first observed in 1827 in grains of pollen in water. After showing that neither currents in the water nor the water's evaporation caused the irregular motion of the pollen grains, the Scottish botanist Robert Brown (1773–1858) supposed that, in this motion, he had discovered the source of vitality common to all forms of life. But upon observing the same irregular motion in particles of fossilized wood, volcanic ash, ground glass, granite, and even a fragment of the Sphinx, Brown gave up this idea.

Investigators following Brown had, by the early 20th century, identified the cause of Brownian motion in the impacts delivered to the microscopic particles, so-called *Brownian particles*, by the molecules composing the surrounding fluid. All that was needed was a quantitative theory of the phenomenon whose predictions could be tested. Albert Einstein provided that theory in 1905. According to Einstein, a group of Brownian particles, all starting from the same point, disperses indifferently in all directions. Also a Brownian particle's mean squared distance $\bar{d}^2$ from its starting point increases as the first power of the time t rather than, as one might expect of uniformly moving particles, as the second power t^2. Brownian particles randomly *diffuse* rather than deterministically *drift*.

In confirming Einstein's theory Perrin constructed a number of diagrams, like the one in figure 63, in which he marked the position of a Brownian particle (often a particle of plant resin) at equal intervals (typically every 30 seconds) and connected successive positions with a straight line. The lengths and numbers of these displacements confirmed the statistical predictions of Einstein's theory. The microscopes with which Brownian particles are seen, in effect, make visible the ordinarily invisible world of atoms and molecules.

However, a common mistake is to associate a single line segment in figure 63, with a single molecular impact. Perrin knew that, if instead of every 30 seconds, he had marked the position of the particle a thousand times more frequently, that is, every 0.03 seconds, he would have created a diagram, except for its size, much like the one above. Every tiny displacement of a Brownian particle hides within itself a pattern of random displacements that mimics the larger pattern. For this reason, the displacements of a Brownian particle are said to be *scale invariant.*

Also in 1905 Einstein created the theory of special relativity and originated the concept of a quantum of light or *photon*. Einstein continued, during the next 20 years of his life, to contribute to the quantum revolution in physics. But eventually he turned his back on quantum theory. In particular, Einstein rejected the probabilistic interpretation of quantum phenomena – an interpretation that rapidly gained ground after Max Born (1882–1970) introduced it in 1926. Yet Einstein began his career by embracing a statistical, that is, a probabilistic, description of Brownian motion. Why then did he reject a probabilistic description of quantum phenomena?

Einstein was comfortable using probability to describe the incompleteness of our knowledge of the natural world. We are ignorant, but neither necessarily nor completely so. According to Einstein, probabilities, properly used, quantify the degree of knowledge and ignorance that follows from our finitude. On the other hand, Max Born's probabilistic interpretation of quantum mechanics radically limits what we can, in principle, know. Einstein rejected Born's use of probability. Rather, such limitation, Einstein believed, simply means that our theories are incomplete. Einstein stubbornly maintained that the fundamental laws of nature (now unknown to us) must be deterministic (not probabilistic or random) and that complete knowledge of an isolated part of the physical world is possible.

In spite of their deep disagreement Max Born and Albert Einstein remained life-long friends. "I at any rate am convinced that He [God] is not playing at dice," Einstein famously wrote to Born in 1926. Many years later Born said of Einstein that,

> He has seen more clearly than anyone before him the statistical background of the laws of physics, and he was a pioneer in the

> struggle for conquering the wilderness of quantum phenomena. Yet later, when out of his own work a synthesis of statistical and quantum principles emerged which seemed to be acceptable to almost all physicists, he kept himself aloof and skeptical. Many of us regard this as a tragedy – for him, as he gropes his way in loneliness, and for us who miss our leader and our standard bearer.

Born went on to say that their disagreement is “based on different experiences in our work and life. But, in spite of this, he remains my beloved master”.

40. Rutherford's Gold Foil Experiment (1910)

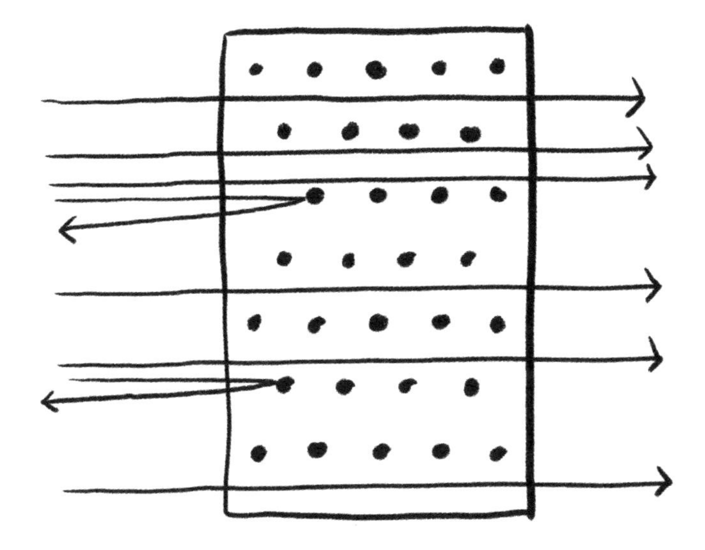

Figure 64

The concept of an atom as a tiny, indivisible building block of the material world is at least as old as the fifth century BCE. Imagine dividing a chunk of matter into smaller and smaller pieces until eventually producing an object that could no longer be divided. This is appropriately an *a-tom*, since the very word means *un-cuttable*. Lucretius, a first century BCE Roman poet, took comfort in the idea that human affairs were mere surface phenomena. At bottom all was "atoms and the void".

Atoms also comforted philosophers because the existence of atoms solved, in part, a philosophical problem. One observes that all things seem to change. But since change is a relative concept, we are moved to ask, "Change with respect to what?" "How can we evaluate change except relative to some unchanging standard that does not itself change?" And so, "How does one account for both change and permanence?" Atoms provide one answer. Atoms are permanent. It is their spatial relationship to one another that changes.

Isaac Newton (1642–1727) endowed Lucretius's invisible atoms with the properties of quite visible objects: mass, weight and the ability to deliver an impact. Daniel Bernoulli (1700–1782) used these Newtonian concepts to explain how and to what extent a gas composed of atoms can exert a pressure on its container walls. But these were speculations, even if essentially correct ones.

The first to call attention to empirical evidence for the existence of atoms was the English chemist John Dalton (1766–1844). Dalton's evidence consisted of the regular proportions of the mass of homogeneous and un-analysable substances or *chemicals* that combine with one another. While Dalton's work reinforced the traditional picture of the atom as a solid, indivisible object, he also brought something new to the idea: each element has its own kind of atom and atoms of the same kind are identical. Dalton's atom remained plausible throughout much of the nineteenth century even as the number of known elements more than doubled from his day to that of the youth of Ernest Rutherford (1871–1937). Rutherford's comment that "I was brought up to look at the atom as a nice hard fellow, red or grey in colour according to taste" must have expressed what many of his generation believed.

But atoms are not so simple. Most importantly, atoms are not even atoms in the sense of being indivisible. For late in the nineteenth century certain kinds of atoms were found to be *radioactive*: that is, to spontaneously emit massive particles or electromagnetic energy or both. Evidently, atoms have parts and some atoms emit their parts: *alpha*, *beta* and *gamma rays* as they were then called. We now know that alpha rays are the nuclei of helium atoms, two protons and two neutrons stuck together; that beta rays are electrons; and that gamma rays are quite short-wavelength, and thus quite high-energy, electromagnetic radiation. Radioactive atoms must in some way contain these various "rays" as constituents.

It was J. J. Thomson (1856–1940), the discoverer of the electron and that beta rays are electrons, who in 1904 quite plausibly suggested that all atoms contain a number of electrons. In order to provide for the charge neutrality of most atoms, Thomson imagined that these atomic electrons reside within a sphere of positively charged, atomic fluid that leaves the entire atom electrically neutral. This model of atomic structure became known as the "plum pudding" or "currant bun" model after edible concoctions of the day. Presumably, the plums or the currants are electrons, while the pudding or the bun dough is the positive fluid that surrounds and neutralizes these electrons.

Thomson's model did not last long. In 1910 Ernest Rutherford and his associate Hans Geiger and student Ernest Marsden bombarded a thin gold foil with the alpha particles (helium nuclei) emitted from a sample of naturally radioactive radium. One day Geiger reported to Rutherford that the gold foil deflected alpha particles back toward their source.

Since an alpha particle is 8,000 times more massive than an electron, an electron residing within an atom could no more deflect an alpha particle from its straight-line path than a fly could deflect a rolling bowling ball. Neither could the whole mass of an atom deflect an alpha particle if, as was supposed, that mass was distributed uniformly throughout the atom's volume. Only if most of the gold atom's mass was concentrated within an essentially point-like core or *nucleus* would a few alpha particles, each directed squarely at a nucleus, bounce back toward their source. Rutherford understood all this and knew what Geiger and Marsden's result implied. Some years later he remarked, "It was quite the most incredible event that has ever happened to me in my life. It was almost as incredible as if you fired a 15-inch shell at a piece of tissue paper and it came back and hit you."

Figure 64 illustrates the essential physics of Rutherford, Geiger, and Marsden's experiment. Several alpha particles approach the gold foil from the left. The foil is represented by an array of gold nuclei shown here as dots. In fact, Rutherford's gold foils were about 4,000 atoms across. Most alpha particles pass through the foil without deflection, but a few bounce directly back.

The orbit of an alpha particle scattering from a gold nucleus is structurally identical to the orbit of a comet approaching, passing around, and then receding from the Sun as illustrated in figure 65. (The alpha particle and nucleus are on the left and the comet and Sun are on the right of the diagram.) Such trajectories have been understood since the time of Newton. Rutherford had only to adapt Newton's general mathematical description to his particular experiment to quantify the observed scattering.

Figure 65

Rutherford's gold foil experiment showed that an atom is composed of a small nucleus and attendant electrons that orbit throughout a much larger volume around that nucleus. A typical nuclear radius is to the radius of its atom as the radius of a beach ball is to the radius of the Earth. Apparently, Lucretius's void is inside as well as outside the atoms that make up our world.

41. X-Rays and Crystals (1912)

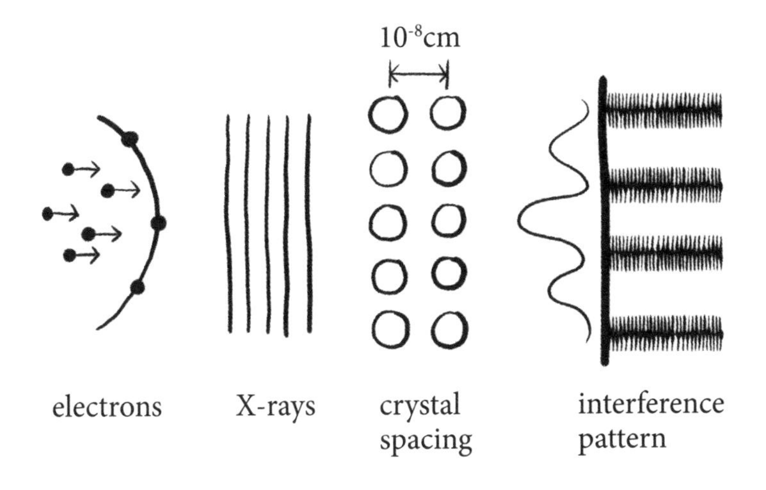

Figure 66

On 8 November 1895, while experimenting with a beam of electrons created within an evacuated glass tube, Wilhelm Conrad Röntgen (1845–1923) accidentally discovered certain "rays" that propagated beyond the end of his tube. These rays seemed to travel in straight lines, made fluorescent materials glow, and exposed photographic plates. Because the rays travelled through flesh but not through bone, Röntgen used them to photograph the bones in his wife's hand. He called the rays *X-rays*.

Röntgen's X-rays were immediately hailed as a new method of photography. The *New York Times* covered Röntgen's discovery early in 1896. That year more than 1,000 professional and popular articles and 50 books and pamphlets were published on X-rays. Röntgen, however, was not pleased with the publicity and complained that, "I could not recognize my own work in the reports". But he had started something new. That spring the young Ernest Rutherford wrote to his fiancée that "every Professor in Europe is now on the warpath" trying to understand X-rays.

That understanding came slowly, but by 1912 evidence was accumulating that X-rays were very high frequency, short-wavelength electromagnetic waves. The left half of figure 66 depicts this understanding. Electrons are accelerated to high speeds and collide with the end of an evacuated glass tube. In the collision, short-wavelength electromagnetic waves – the X-rays – are created that carry forward the energy and momentum of the electrons. Yet not everyone was convinced. Some continued to believe that X-rays were particles.

Max von Laue (1879–1960), a near contemporary and friend of Albert Einstein, proposed an experiment (shown in the right half of figure 66) whose result secured the case for waves. Early in 1912, while listening to a student explain his research on the interaction of long-wavelength electromagnetic waves with the atoms or molecules that compose a crystal, von Laue asked himself, "Why not shine X-rays on a crystal?"

Since the spacing between the atoms or molecules in a typical crystal (10^{-8} cm) is only a little larger than the estimated wavelength of X-rays (10^{-9} cm), X-ray waves should, after passing through the crystal, produce an *interference pattern*: that is, a pattern of constructively and destructively superposing waves. This interference pattern should be similar to that produced by visible light passing through a regular series of parallel slit-shaped openings called a *diffraction grating*. In both cases, the interference pattern produced depends upon a wave property called *diffraction*: that is, a departure from straight-line propagation.

Although X-ray interference is, in its geometry, a smaller-scale version of visible light interference, physically the two cases are quite different. X-rays pass through a crystal by vibrating the charged particles in the atoms (or molecules) composing the crystal. These atoms, in turn, radiate new waves that pass on the interaction from atom to atom until the very last atoms on the far side of the crystal radiate like a string of regularly spaced radio beacons. Visible light, on the other hand, passes freely through the slits of a diffraction grating and is absorbed or reflected by the material surrounding the slits.

Laue convinced two colleagues, Walter Friedrich and Paul Knipping, to test his idea. Their initial experiment, with materials and equipment on hand, captured on film the X-ray interference pattern shown in figure 67. It consists of several dark splotches, each indicating the constructive interference of diffracted X-rays, surrounding a single, larger dark patch, indicating the remains of the original ray. This image attracted favourable attention and secured funding for more refined experiments that fully confirmed von Laue's detailed analysis. Laue, Friedrich and Knipping published their first results in June of 1912.

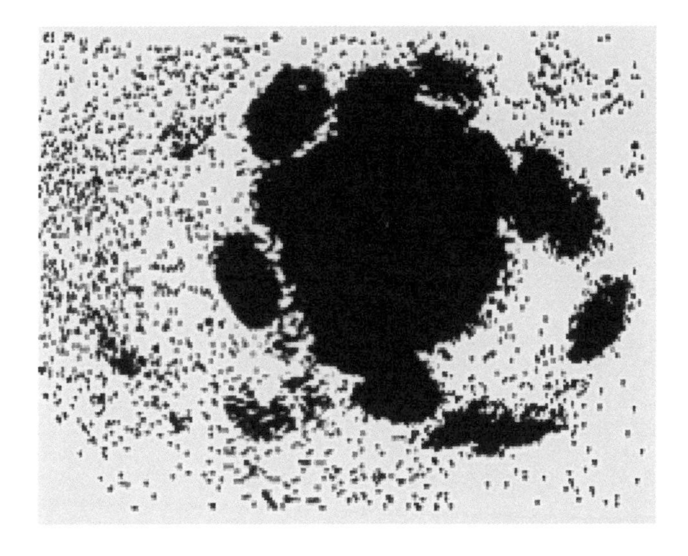

Figure 67

Laue's idea was brilliant and its confirmation complete. In place of pursuing a single idea for many years, he had "suddenly ... perceived the way which subsequently proved to be the shortest path to success". The Nobel committee awarded Von Laue its 1914 prize in physics for "his discovery of the diffraction of X-rays by crystals". Laue's rise, from *privatdozent* (junior faculty with no regular salary) to Nobel Laureate, took less than three years.

Laue survived long enough to be tested by the fires of Nazism and World War II. He spoke out publicly against the persecution of the Jews and the promotion of a "German science" that, for instance, rejected relativity because Einstein was Jewish. He remained in Germany during the war, an outspoken critic of the Nazis, secretly helping his Jewish colleagues to emigrate and then to escape. After the war Laue helped rebuild Germany's institutions of science. Then, in 1960, a motorcycle struck and overturned the car he was driving to work. In the few days left to him Laue composed his own epitaph: "He died trusting in God's mercy."

42. Bohr's Hydrogen Atom (1913)

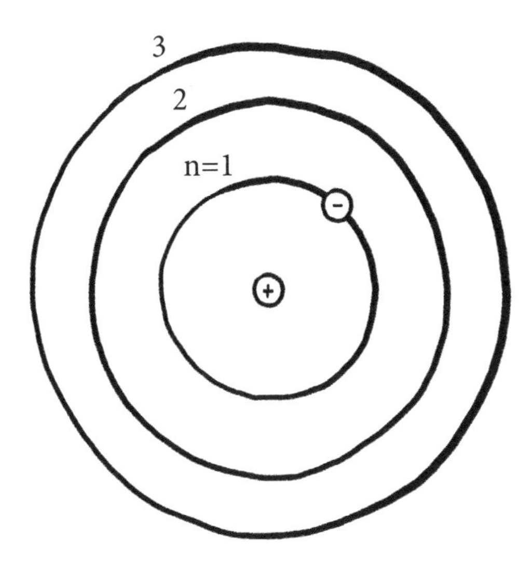

Figure 68

The Danish public knew Niels Bohr (1885–1962) as a soccer player before it knew him as a physicist. Crucial to the evolution of Bohr's persona from sports hero to Nobel laureate was the postdoctoral year (1911–1912) he spent in England studying atomic physics, first with J. J. Thomson at the University of Cambridge and then with Ernest Rutherford at the University of Manchester. Recall that the earlier work of Max Planck (1900) and Albert Einstein (1905) had suggested that the concepts of classical physics were not sufficient for understanding the atom.

Bohr set for himself the task of understanding the simplest of all atoms: the hydrogen atom. Initially he was uncertain about how to proceed. Rutherford's gold foil experiment had implied that most of the mass of an atom resides in a tiny, positive core or nucleus. And because electrical and gravitational forces were similarly structured it was natural for Bohr and others to suppose that the electron in a hydrogen atom moves round its nuclear proton in a circular or elliptical orbit just as a planet moves around the Sun. The problem with this planetary picture of the atom is simple and dramatic. According to classical concepts an electron that orbits a nucleus will radiate for the same reason that an electron moving back and forth along a radio antenna will radiate. Both are accelerating and, therefore, both radiate energy. Consequently, an atomic electron will lose energy and spiral into the nucleus. The atom will quickly collapse.

Yet stable atoms do exist. And by 1912 scientists knew that the diameter of a hydrogen atom was about 10^{-8} centimetres. Furthermore, the frequencies and wavelengths of the light that could be absorbed and emitted by atoms of different elements were known quite precisely and quantified in certain tantalizing, simple, yet unexplained formulas. Any successful model of the structure of the hydrogen atom would have to be consistent with these well-established facts.

Bohr's working principle was to make the least modification to the classical physics of the hydrogen atom needed in order to account for its stability and its interaction with light waves. He implemented this principle by simply asserting that the hydrogen atom observes classical physics except that it does not radiate when its electron occupies one of a discrete set of special orbits. He called these special orbits *stationary states*. Bohr's assertion, though quite unjustified, worked brilliantly.

Of course, in order to quantify his model, Bohr had to specify what makes an orbit a stationary state. He did this by requiring that in a stationary-state circular orbit the electron's angular momentum – that is, the product of its radius, its mass and its speed – is a multiple of the fundamental constant discovered by Max Planck during his study of blackbody radiation and now called Planck's constant.

In order to illustrate the consequences of these assumptions we number the allowed stationary-state circular orbits with an index n=1,2,3 . . . , as illustrated in figure 68, so that, for example, E_1 and r_1 stand for the energy and radius of the first, most stable, *ground,* stationary-state orbit. In general the electron's energy E_n and its distance from the nucleus r_n in a stationary state increase with increasing index. Only the three innermost, circular orbits of a hydrogen atom's electron are shown in figure 68, but possible orbits extend out in ever larger, unevenly spaced, concentric circles.

According to Bohr, the hydrogen atom can absorb energy from light only by boosting its electron from a less energetic, lower, stationary state orbit into a more energetic, higher one – as shown in figure 69 – and likewise can emit light energy only when an electron drops from a more energetic, higher, stationary state orbit into a less energetic, lower one. This hypothesis allowed Bohr to calculate the frequency of the light absorbed or emitted when the electron transitioned from one orbit to another – frequencies that exactly reproduced those observed.

Bohr's model was quickly recognized as important. Rutherford's assessment was typical, "While it is too early to say whether the theories of Bohr are valid, his contributions . . . are of great importance and interest." But Bohr and his contemporaries had less success in applying the model assumptions to multi-electron atoms. This failure should have been expected. For Bohr's method depended upon using as much of classical physics as possible in describing an electron orbit. However, helium, the next simplest atom after hydrogen, is composed of three particles: one nucleus and two electrons. And unlike the classical two-body problem, the classical three-body problem cannot be exactly solved. The quantum revolution of the 1920s dismantled the very notion of an orbit of an atomic electron. Still, in 1913, the simplicity and success of Bohr's model of the hydrogen atom forced physicists to take a close look at his model and ask themselves, “Why does it work so well?”

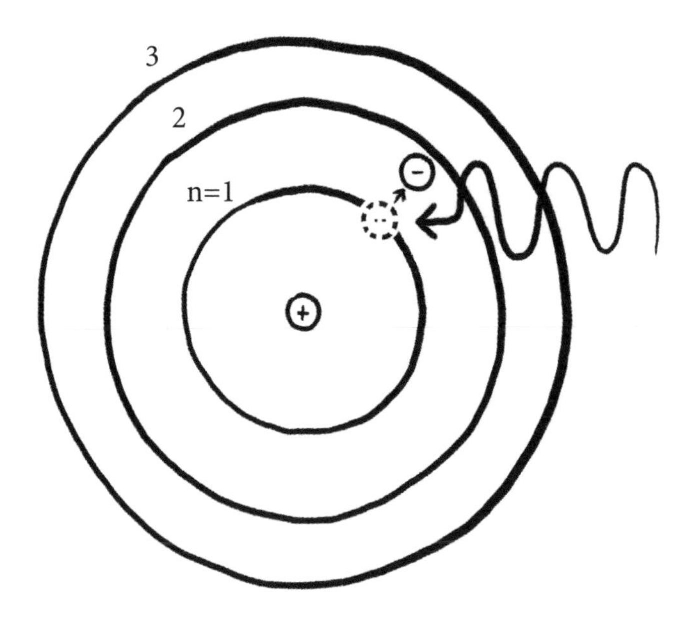

Figure 69

43. General Relativity (1915)

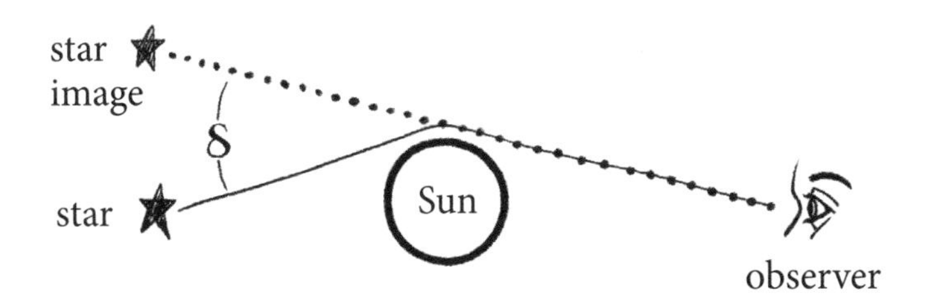

Figure 70

Each side in the Great War of 1914–18 tried to bleed the other to death. Because newly developed machine guns and heavy artillery made traditional offensive tactics obsolete, furious battles achieved little or nothing of military value – only massive death. The Battle of Verdun, for instance, lasted nine months, created a million casualties, and left two depleted armies occupying much the same ground as before. By the time Germany sued for peace in the autumn of 1918, a whole generation of "doomed youth", in Wilfred Owen's haunting words, had died "as cattle".

The memory of the Great War – its chauvinism, its horror and its futility – was still fresh on 6 November 1919, one year after the armistice, as the Royal Society and Royal Astronomical Society convened in London to announce observations that confirmed Einstein's theory of general relativity. That English scientists had made the considerable effort necessary to test an esoteric theory of a German-speaking Swiss scientist was welcome news to a public weary of war.

The English scientists tested Einstein's theory by placing themselves within the shadow cast by the Moon as it passed in front of the Sun during the total eclipse of May 29, 1919. On that date the oval-shaped shadow of total eclipse, some 100 miles across, travelled from the east coast of Brazil to the west coast of Africa. One team of observers led by the Royal Astronomer Andrew Crommelin was stationed at Sobral, Brazil and the other led by the Cambridge physicist Arthur Eddington was stationed on Principe Island off the west coast of Africa, near present-day Gabon. Each team measured the angular separation of two stars as one of them passed near the edge of the obscured Sun's disc. When this separation was compared with earlier measurements of the same separation, the position of the star grazing the Sun's disc during total eclipse had shifted as if the Sun had attracted its light – illustrated in figure 70 in exaggerated form.

That the Sun can attract starlight was implicit in Newton's theory of gravitation and his proposal that light is composed of tiny, massive particles travelling at high speed. Accordingly, Newton could have calculated the angle δ through which starlight is deflected in the circumstance illustrated above, but he did not. The first to do so, according to Newtonian principles, was Johann Georg von Soldner (1776–1833) who in 1801 predicted a deflection of 0.87 seconds of arc – somewhat less than 1/3600th of a single degree of arc. [Note: one degree of arc is 1/360 of a complete circle.]

Einstein, like Newton, believed that light is composed of tiny, massive particles. A theory based simply on tiny, massive particles of light (however differently conceived by Newton and Einstein) and Newton's theory of gravitational attraction leads to Soldner's deflection of 0.87 seconds. However, something more is at work in the general relativistic description of space, time and gravitation that Einstein proposed in 1915. That additional something is the idea that massive bodies curve space and time in their vicinity. When this curvature is accounted for, Einstein's theory predicts a deflection of 1.74 seconds of arc – exactly twice that of Soldner's prediction.

Crommelin and Eddington confirmed Einstein's general relativistic prediction, not Soldner's Newtonian one. Shortly after the public announcement of their results, *The Times* of London boldly declared that Einstein had "overthrown" Newtonian physics and the *New York Times* declared that he had "knocked out" Euclidean geometry. The cover of the popular *Berliner Illustrirte Zeitung* displayed a full-page photograph of a thoughtful Einstein. J. J. Thomson (1856–1940), the discoverer of the electron, pronounced that "his [Einstein's] was one of the greatest achievements of human thought". Such coverage turned Einstein, who was already well known among physicists, into an icon of science.

That some of the consequences of the special and general theories of relativity are counter-intuitive has contributed to Einstein's long-lasting fame. Recall that general relativity sprang from special relativity – the latter a theory of moving clocks that run slow and moving metre sticks that shrink. Special relativity was and is a spectacular success. The daily, predictable operation of thousands of particle accelerators around the world attests to its correctness. Its generalization, the general theory of relativity, has been confirmed in more limited circumstances. Besides the deflection of starlight, general relativity explains, in part, the slow advance of Mercury's closest approach to the Sun and also the red shift of light emerging from massive bodies and the properties of black holes.

Yet Einstein did not set out to predict unusual phenomena. His motivation, according to the eminent physicist Subrahmanyan Chandrasekhar (1910–1995), was more aesthetic than empirical. His desire was for simplicity, system and symmetry in physical theory rather than for successfully accounting for specific phenomena. Of course Einstein looked for and found applications of his theories, but in 1915 general relativity satisfied no outstanding empirical need.

Interestingly, what counts as a good theory has changed over time. Today general relativity has an odd reputation among physicists. While its successes cannot be denied, many are uncomfortable with the absence of a quantum theory of gravity: that is, uncomfortable with the absence of a quantum version of general relativity. This absence is felt more as a lack of conformity to current standards than as any failure of general relativity's predictions. For the language of quantum mechanics has become the language of physics. And gravity alone among fundamental forces has resisted expression in that language.

44. Compton Scattering (1923)

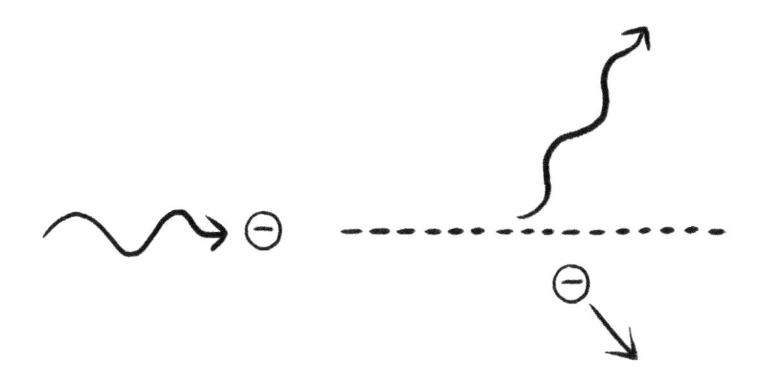

Figure 71

Einstein invented the concept of light quanta in 1905 in order to account for the photoelectric effect: that is, for the capacity of ultraviolet light to eject electrons from metallic surfaces. According to his hypothesis, the energy in light is concentrated in bundles or quanta that are, on occasion, entirely and instantaneously transferred to a single electron. The energy $E[=h\nu]$ in each quantum of light is determined by the frequency of the light wave ν with which it is associated where the proportionality constant h is *Planck's constant.*

Einstein's explanation of the photoelectric effect was not experimentally confirmed until ten years after its proposal. Even then most physicists continued to resist the implication of Einstein's explanation that light consists of quanta or *photons,* as they would later be called. After all, the diffraction and interference of visible light only makes sense if light is composed of waves. Infrared, ultraviolet and X-ray radiation are also waves – straightforward extensions of the visible spectrum. Furthermore, the wave theory of light is firmly grounded in Maxwell's theory of electromagnetism – a theory so successful it could not be questioned. The few experiments that light waves did not explain concerned the interaction of light with atoms and molecules – an interaction that, because it was not fully understood, could be ignored by quantum sceptics.

Max Planck probably spoke for many when, in 1913, he signed a letter proposing Einstein for membership in the Prussian Academy of Sciences.

> In sum, one can say that there is hardly one among the great problems in which modern physics is so rich to which Einstein has not made a remarkable contribution. That he may have sometimes missed the target in his speculations, as, for example, in his hypothesis of light quanta, cannot really be held too much against him, for it is not possible to introduce new ideas even in the most exact sciences without sometimes taking a risk.

Einstein himself considered light quanta as a mere provisional device that in time would be replaced with a more foundational theory.

In the meantime Einstein used the concept of light quanta cautiously. He spoke, for instance, of the ejection of photoelectrons as akin to drawing beer from a barrel. That beer is always drawn from barrels in pint containers does not mean the beer inside the barrel is partitioned into pints. In similar fashion, that light, in some circumstances, seems to give up its energy in standardized chunks does not mean that light is composed of quanta.

It fell to Arthur Holly Compton (1882–1962) to endow light quanta with a reality that few could question. Figure 71 illustrates Compton's experiment. Light (on the left) is directed at a free electron (represented by the leftmost circle). Light (on the right) scatters from the electron and the electron recoils. The dotted line shows a continuation of the electron's original trajectory. According to Compton, the angle through which the light scatters and its shift to lower frequency and longer wavelength is related to the direction of the recoiling electron and its kinetic energy exactly as if the energy and momentum of the light were concentrated in a single quantum. The light quanta and an electron collide, say, as one billiard ball with another.

In fact, Compton scattered X-rays, instead of visible light, from the weakly bound electrons in the carbon atoms of a sample of graphite, instead of from perfectly free electrons. Recall that William Röntgen had discovered X-rays in 1895 and that Max von Laue had demonstrated in 1912 that X-rays are relatively high frequency (compared to visible light) electromagnetic waves. Now Compton had shown that X-rays, and by extension all parts of the electromagnetic spectrum, also behave as particles of electromagnetic radiation.

Interestingly the 1923 paper in which Compton announced his result did not mention Einstein's 1905 paper on the photoelectric effect. Yet Compton's experiment is routinely characterized, as we do here, as confirming Einstein's light quanta hypothesis. Compton's biographer, Roger Steuwer, claims that Compton, given the way he typically mischaracterized Einstein's work, "quite likely never even read Einstein's 1905 paper". Steuwer continues, "One of the most striking aspects of Compton's research program, when viewed in its entirety, is its relative autonomy", that is, ". . . his theoretical insights were derived from, and anchored in, his own precise experiments."

The *Compton effect* created a sensation. Before Compton's experiment and its interpretation one could confine the supposed particle-like qualities of light to its ill-understood interaction with matter. After Compton's experiment, physicists had no choice but to embrace light quanta.

But the Compton effect created a new difficulty. Two incompatible theories of light, a particle theory and a wave theory, had been found indispensable, each in different circumstances. High-frequency light interacts with electrons just as if the light were composed of photons. Yet beams of low-frequency light also create interference patterns just as do waves. Light can be treated either as a bundle of energy E (=$h\nu$) or as a wave with frequency ν (=E/h). Eventually physicists became accustomed to this *wave-particle duality*, and the idea remains useful to this day. But a more coherent theory has now been fashioned. Shortly after World War II the wave and particle theories of light were integrated into a single mathematical theory called *quantum electrodynamics*.

45. Matter Waves (1924)

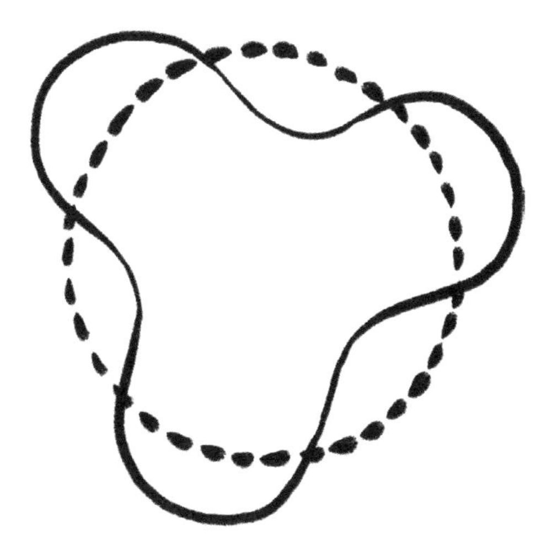

Figure 72

Louis de Broglie (1892–1987) (pronounced *Louie de Broy*) was born into an illustrious family that had since the 17th century produced prominent soldiers, politicians and diplomats for France. Louis's father was the fifth duc de Broglie. Louis would eventually become the seventh. Educated as a child by private tutors, he completed high school and matriculated at the University of Paris, first studying history, then law, and finally physics – especially theoretical physics. Although World War I interrupted his studies, his older brother, a prominent experimental physicist, arranged for Louis to be posted, for much of the war, to the safety of a telegraph station at the foot of the Eiffel Tower. Demobilized in 1919, Louis returned to the University in order to finish his doctoral dissertation.

A confidence born of his family's privilege and means and his obvious talent for physics enabled Louis to develop ideas outside the mainstream of physics. Einstein's 1905 work on relativity and on the photoelectric effect inspired de Broglie. In particular, Einstein's idea that light, so successfully described as a wave, could also behave as a quantum of energy suggested to de Broglie the complementary idea that an electron, considered a particle since its discovery in 1897, could also behave as a wave.

According to de Broglie's hypothesis, every particle is accompanied by a wave that helps determine its behaviour. De Broglie found that the wavelength λ of the wave associated with a material particle, today called the *de Broglie wavelength*, is inversely proportional to the particle's momentum p so that $\lambda = h/p$ where h is Planck's constant. The smaller the particle momentum, the larger its wavelength, and the larger its wavelength, the more evident the particle's wave nature. These wave properties were, to de Broglie, as real as the mass of the associated particle.

De Broglie's first application of matter waves was to the hydrogen atom. Niels Bohr (1885–1962) had, in 1913, somewhat arbitrarily limited the continuum of its possible electron orbits to a discrete set he called *stationary states*. According to Bohr, only an electron in a stationary state orbit is stable and immune from the classical expectation of radiating energy. Bohr's model was empirically successful even if based on *ad hoc* postulates.

De Broglie's matter waves provided an explanation of Bohr's arbitrary limitation of electron orbits that, in turn, opened up a new world of thought. He found that only electron orbits associated with waves that smoothly reconnect with themselves after a complete orbit are possible. Waves associated with other orbits destroy themselves. De Broglie's smoothly reconnecting waves neatly correspond to Bohr's stationary states. Figure 72, more conceptual than representational, of a circular electron orbit associated with a wave of three complete wavelengths illustrates this concept. Other possible orbits are associated with waves of a whole number (1, 2, 3,. . .) of wavelengths.

The idea of a wave constructively interfering with itself may already be familiar. Fix one end of an extendable, elastic cord to a post. (A Slinky in place of a cord will do.) Extend the cord, and launch waves by moving its free end up and down. Only those regular up and down motions with certain discrete frequencies will produce waves that constructively interfere with the waves reflected from the fixed end of the cord. With this particular setup the only large waveforms possible are those composed of a whole number of half wavelengths. Figure 73 shows a cord, fixed at one end, supporting a wave of three half wavelengths.

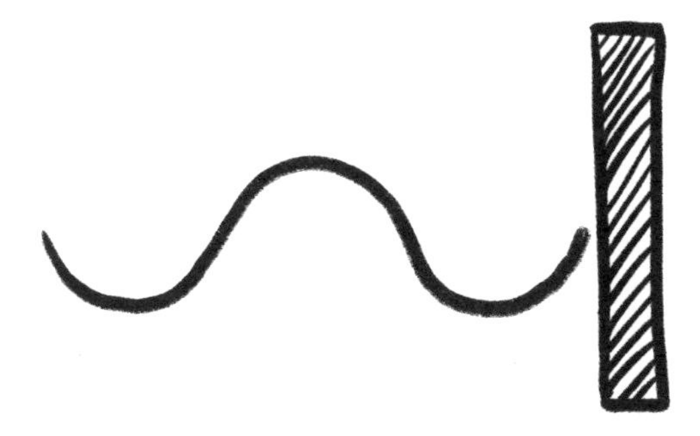

Figure 73

De Broglie composed his ideas on matter waves into a doctoral dissertation in 1924. Upon reading it Einstein claimed that, "He [de Broglie] has lifted a corner of the great veil." Erwin Schrödinger (1887–1961), by developing a wave equation whose solutions described and generalized de Broglie's matter waves, soon lifted another corner of that veil.

Experimental confirmation came quickly. Indeed, the American physicist Clinton Davisson (1881–1958) confirmed de Broglie's matter waves even before he knew he was doing so. Davisson and his collaborator, Lester Germer, had been continuing Davisson's earlier study of the surfaces of crystals by shining a beam of low-energy electrons on those surfaces and recording the intensity of the reflected beam as a function of its angle of incidence. Germer, on a trip to Europe in 1926, was astonished to hear a lecture in which the presenter, Max Born (1882–1970) – who a year later developed the probabilistic interpretation of matter waves – showed a curve from Davisson's earlier research that, Born claimed, showed that electrons reflect from a crystal's surface just as do waves. On Germer's return he and Davisson refined their experiment in the light of Born's comment. Their result: slow electrons reflect from the surface of a crystal exactly as do waves with a wavelength equal to that of the de Broglie wavelength λ [$=h/p$] where p is the electron momentum.

De Broglie received the Nobel Prize in physics in 1929 "for his discovery of the wave nature of electrons". Erwin Schrödinger received the prize in 1933, Clinton Davisson in 1937, and Max Born in 1954. De Broglie, like Einstein and Schrödinger, rejected Born's probabilistic interpretation of matter waves. They were too strongly committed to the classical tradition of continuity and determinism in physics. Yet Born's quantum probabilistic interpretation has proven all but inescapable.

46. The Expanding Universe (1927–29)

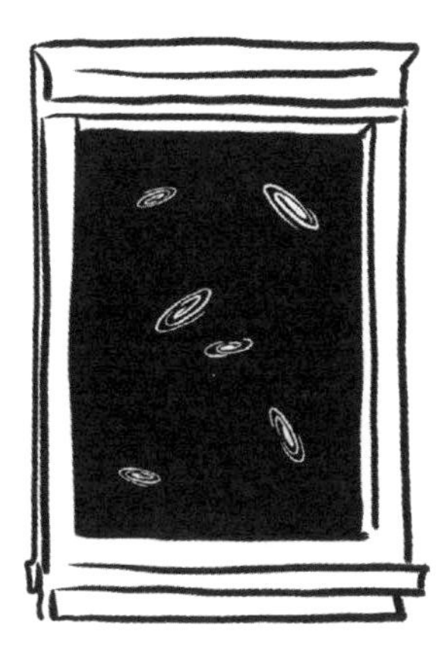

Figure 74

The daily appearances of Sun and Moon, the moving planets, the starry sky, and the encircling band of hazy or "milky" light called the *Milky Way* have, over the centuries, invited men and women to consider the universe as a whole. Of what is it composed? Does it move? What is its shape? are questions they asked and sometimes answered. Observations made with the naked eye or with the aid of a telescope only slightly constrained their speculations. Experiments were, of course, out of the question.

Immanuel Kant (1724–1804) was one inquirer whose speculations were disciplined with reason. He reasoned both from what little he knew (the Sun is part of the Milky Way star system) and from what he could reasonably suppose (the laws of physics are everywhere the same) to a cosmology much in advance of his time. According to Kant the Milky Way must be an extended rotating system of stars whose apparent stability is the result of a balance between the attractive force of gravity and the centrifugal tendency of rotating systems to fly apart. While we see the Milky Way from its inside, a distant observer on its outside would see a hazy, flattened ellipse that appeared much like the nebulae of unknown composition astronomers had been discovering in Kant's day. Thus, it was likely that the Milky Way was not the only such star system – a lone island in an otherwise empty universe – but one among many similar systems scattered throughout the universe.

Empirically minded astronomers ignored Kant's reasoning, based as it was on a slim foundation of physical evidence. More influential were the painstaking telescopic studies of William Herschel (1738–1822) and Harlow Shapley (1885–1972). Shapley determined the size of the Milky Way by using the characteristic relation, discovered by Henrietta Swan Leavitt (1868–1921) (one of the first female "computers" at the Harvard College Observatory), between the brightness of the variable stars called *Cepheids* and their period of variation. Shapley developed a method of determining the distance of a particular Cepheid variable by observing its period and consequently inferring its absolute brightness, and then by comparing the latter to its apparent brightness, inferring its distance. Since Cepheid variables are scattered throughout the Milky Way, Shapley was able to determine the Milky Way's size – some 100,000 light years across. However, in opposition to Kant, he erroneously concluded that the nebulae were objects in the Milky Way and that the Milky Way encompassed all the matter of the universe.

This last conclusion fell apart in 1923–24 when Edwin Hubble (1889–1953), using the recently constructed 100-inch diameter optical telescope on Mount Wilson in California, identified Cepheid variables in the Andromeda nebula. Using Shapley's method of determining distances, Hubble found that this nebula was, in fact, a separate star system some ten times further from the Milky Way than the latter is across. Upon reading a letter from Hubble explaining this discovery, Shapley remarked to a colleague, "Here is the letter that destroyed my universe."

Hubble went on to study the spectra: that is, the characteristic pattern of coloured light emitted from and absorbed by the gases in the atmosphere of the stars that compose distant nebulae. Interestingly, Hubble found that, more often than not, these spectra were shifted toward longer wavelengths and lower frequencies ("red-shifted") relative to the spectra of the same gases in terrestrial laboratories. Hubble then supposed these red shifts were *Doppler shifts* – a consequence of a galaxy's rapid recession from our galaxy, just as the pitch of the siren of an emergency vehicle is lower (and its sound wavelength longer) when the vehicle rapidly recedes from the listener. He found that the further the galaxy, the faster its velocity of recession in direct proportion one to the other – a relation now known as *Hubble's law*.

Milton Humason (1891–1972), a local Mount Wilson mule-driver and observatory janitor who turned himself into a meticulous and able astronomer, assisted Hubble in this work. Their discovery at first glance seemed to place our galaxy, the Milky Way galaxy, at the centre of a giant cosmic explosion. After all, the most quickly moving fragments produced in an explosion would travel, in a given interval, furthest from its centre. But another idea has prevailed. All modern cosmologies assume that, on the largest scales, the matter of the universe is uniformly distributed. Thus, the distribution of galaxies looks the same in different directions and would look the same in different places. According to this assumption, called the *cosmological principle*, there is no centre and no edge of the universe. The "cosmic explosion" happened everywhere at the same time.

Hubble's law and the cosmological principle together imply that the density of galaxies decreases as time marches on. Figure 74 illustrates this conclusion by imaging two views of the cosmos through the frame of a single window. The right-hand view is billions of years later than the left-hand view and consequently shows a universe less densely populated with galaxies. It is in this sense that we can say the universe is *expanding*.

There is much evidence supporting this universal expansion. However, the rate of expansion and its proximate cause are still (in 2017) under investigation. Although Hubble was the first to gather data that showed galactic redshifts, he was not the first to interpret that data in terms of a universal expansion. Rather, that honour is due to Georges Lemaître (1894–1966), a Belgian Roman Catholic priest and student of Arthur Eddington, the latter one of the leading general relativists of his time. Lemaître discovered a solution to the equations of general relativity that describes an expanding, homogeneous, and isotropic universe with non-zero density and found evidence for this solution in Hubble's earliest data. Published in a little-known journal in 1927, Lemaître's work, eventually praised by Einstein and others, went, for a time, unnoticed.

Hubble eventually took the unusual step of hiring an agent to promote his case for a Nobel Prize in physics. This was an uphill and ultimately unsuccessful battle – for in Hubble's day astronomers were not considered for Nobel Prizes. Even so Hubble deserves recognition at the level of a Nobel Prize for his discovery that the nebulae are independent star systems and for his crucial role in discovering the expansion of the universe.

47. The Neutrino and Conservation of Energy (1930)

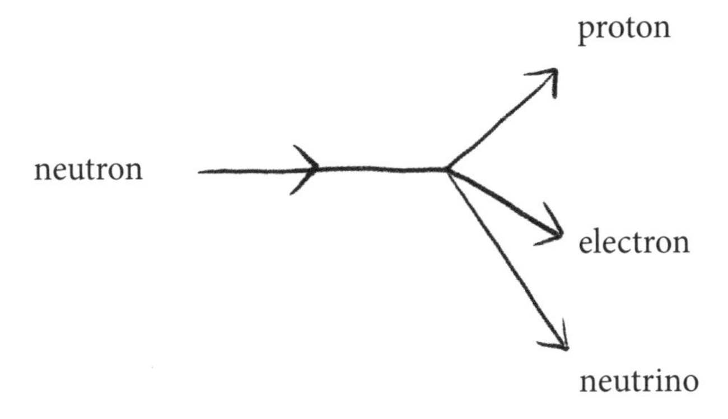

Figure 75

When first discovered in the last few years of the nineteenth century, radioactivity troubled physicists. After all, atoms were supposed to be indestructible. How then could they eject their parts? And why did they even have parts? Furthermore, radioactivity seemed to be an inexhaustible source of energy. Whence came this energy? From the atom itself, the region surrounding the atom, or was it created *ex nihilo* at the moment of radioactive decay?

Hard work and imagination soon answered these questions. Atoms, indeed, have parts. Radioactivity consists of unstable atoms randomly ejecting those parts in alpha, beta and gamma radiation. Alpha (helium nuclei) and beta (electron) radiations transform the original atom into one of another kind. Ernest Rutherford summarized the situation in 1904:

> This theory [of transformation] is found to account in a satisfactory way for all the known facts of radioactivity and a mass of disconnected facts into one homogeneous whole. On this view the continuous emission of energy from the active bodies is derived from the internal energy inherent in the atom, and does not in any way contradict the law of conservation of energy.

Einstein's discovery, in 1905, that anything with mass *m* has energy *E* in the amount mc^2 and anything with energy *E* has mass *m* in the amount E/c^2 supported Rutherford's assessment. (Here *c* is the speed of light.) The energy of radiation is accounted for by a loss of the radioactive atom's mass.

By 1929 physicists had made more progress. Rutherford discovered the nucleus in 1910, and Werner Heisenberg, Erwin Schrödinger and Paul Dirac fashioned different approaches to the quantum mechanics of atomic structure (respectively in 1924, 1926 and 1928). Soon similar ideas would be applied to the nucleus. According to Dirac:

> The underlying physical laws necessary for the mathematical theory of a large part of physics and the whole of chemistry are thus completely known, and the difficulty is only that the application of these laws leads to equations much too complicated to be soluble.

And yet, problems remained whose solution required more than "the application of these laws".

The electrons ejected from unstable nuclei in beta radiation were at the heart of these problems. In particular, the electron energies in beta radiation are continuously distributed over a range of possible values even when their source is a single kind of nucleus. In contrast, alpha and gamma radiation – respectively, helium nuclei and high-frequency electromagnetic waves – have well-defined energies that characterize the particular nuclei from which they come.

In 1929 physicists believed (correctly) that nuclei are composed of protons and neutrons in sufficient numbers to account for their nuclear charge and mass. Many also believed (incorrectly) that a neutron is a tightly bound system of a proton and an electron. Because a proton and electron are oppositely charged they attract each other. For this reason these scientists thought that beta rays are produced when a neutron breaks in two and releases enough energy to drive its proton and electron apart.

As reasonable as this picture might seem, it is unacceptable. For given that both momentum and energy are conserved in any process and that the neutron is slightly more massive than the proton (each about 1,800 times more massive than an electron), the electron must carry away almost all the energy released in the beta decay of a neutron. Furthermore, betas, like alphas and gammas, should have well-defined energies that characterize the neutron from which they are emitted. But, in fact, the energy of the emitted electrons varies widely. Either identical neutrons contain different amounts of energy or, as suggested by Niels Bohr, energy is not conserved in a single decay but only on average over many decays. Most physicists rejected these possibilities.

Wolfgang Pauli (1900–1958) rescued the situation, in 1930, by suggesting that the energy released in the beta decay of a neutron is shared among the remaining proton, the electron, and an as yet undetected, lightweight particle – later named a *neutrino* (Italian for "little neutron") by Enrico Fermi (1901–1954). Figure 75 illustrates the decay of a neutron into a proton, an electron and a neutrino. Because the electron sometimes takes more and sometimes less of the available energy with most of the rest going to the neutrino, Pauli's version of beta decay conserves energy while allowing for a continuous distribution of electron energies.

That the neutrino interacts only weakly with other particles explains why it had not, in Pauli's time, been detected. Pauli's proposal took great courage. Privately he confided to a friend, "I have done something very bad today by proposing a particle that cannot be detected; it is something no theorist should ever do." But Pauli started something of a trend. Today we have particle physicists whose job it is to suggest new particles every time an unusual experimental result is observed.

The Italian physicist Enrico Fermi took Pauli's idea and, using the newly developing methods of quantum electrodynamics, constructed a theory of beta decay that predicted the observed distribution of electron energy. According to Fermi's theory, a proton, electron and neutrino are not constituents of the neutron but, rather, are created at the moment of its decay. Fermi's theory, published in 1934, was a great success.

The neutrino is, indeed, difficult but not impossible to detect. In 1956, more than 25 years after Pauli's original proposal, Clyde Cowan and Frederick Reines were confident that they had observed neutrinos produced in a nuclear reactor and sent Pauli the following exciting, if plainly phrased, confirmation: "We are happy to inform you that we have definitely detected neutrinos."

48. Discovering the Neutron (1932)

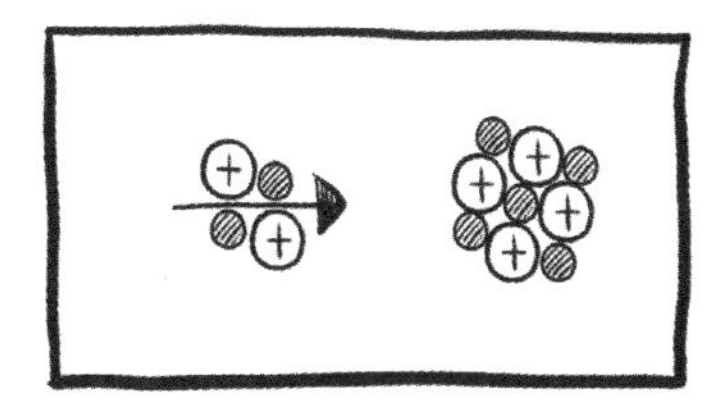
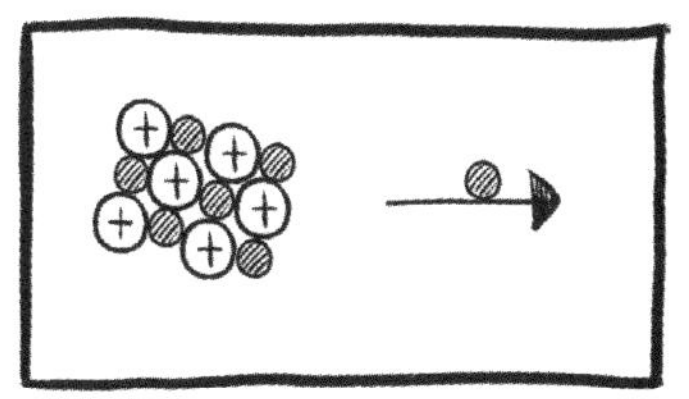

Figure 76

James Chadwick (1891–1974) gets credit for "discovering the neutron", but his actual contribution is not so simply described. He was neither the first to predict the existence of the neutron nor the first to find evidence for its existence. Nor was he the first to realize that neutrons must be elementary rather than composite particles.

Of course, sometimes the word *discover* is quite appropriate. Ernest Rutherford, for instance, certainly discovered the atomic nucleus. Its existence was unsuspected in 1910 when Rutherford, his associate Hans Geiger, and their undergraduate student Ernest Marsden performed an experiment whose inescapable interpretation was that most of the mass of the atom and all of its positive charge were concentrated in a relatively small body at the centre of the atom. Rutherford later called this body a *nucleus.*

Since the hydrogen nucleus is the least massive and least charged of the known nuclei, Rutherford took it to be a building block out of which other nuclei were composed and gave it the name *proton,* after the Greek for "first thing". (Rutherford was an adept eponymist. He not only gave us *nucleus* and *proton* but also *neutron* and *alpha*, *beta* and *gamma* radiation – all names that have survived to this day.)

It soon became apparent that while the number of protons in an atomic nucleus determines an atom's *chemical* properties, a nucleus composed of protons alone does not explain its *physical* properties – in particular its mass. For instance, the next least massive atom after hydrogen is helium. Its nucleus has two protons but weighs approximately four times more than a single proton. Other nuclei weigh at least twice and often more than twice the weight of the protons they contain. Rutherford's solution was to postulate the existence of a neutral particle in nuclei, the *neutron*, with mass approximately equal to that of a proton. Neutrons inhabit nuclei in numbers that bring each nuclear mass up to its observed value. Thus a helium nucleus contains two protons and two neutrons bound together with the strong nuclear force. So far, so good. But Rutherford mistakenly conceived the neutron itself to be a composite system containing a proton and an electron. (The electron mass is about 1/1800 of a proton mass.)

Such was the common understanding and misunderstanding in 1928 when three teams of researchers – Rutherford and his assistant, James Chadwick, in Manchester, England; Walther Bothe and his student, Herbert Becker, in Berlin; and Irène Curie (Marie Curie's daughter) and her husband, Frédéric Joliet, in Paris – began bombarding various light elements with alpha particles: that is, with helium nuclei. These teams found that when a nucleus absorbs an alpha particle it typically becomes unstable and emits, uniformly in all directions, penetrating, high-frequency electromagnetic – that is, gamma – radiation. But when beryllium, with a charge of 4 protons and a mass of 9 atomic mass units, was bombarded, something unusual happened: the unstable nucleus emitted radiation only in the forward direction – that is, in the direction of the bombarding alpha particles.

Chadwick was the first to propose that when a beryllium nucleus absorbs an alpha particle it emits not electromagnetic radiation but another particle, a neutron, in the forward direction, and set about devising an experiment to confirm this proposal. By February of 1932 Chadwick was confident enough to submit a letter to the journal *Nature* entitled "Possible Existence of a Neutron". As Chadwick later explained,

> The results, and others I have obtained in the course of this work, are very difficult to explain on the assumption that the radiation from beryllium is a quantum radiation, if energy and momentum are to be conserved in the collision. The difficulties disappear, however, if it is assumed that the radiation consists of particles of mass 1 and charge 0, or neutrons.

Figure 76 illustrates Chadwick's idea. In the left frame an energetic alpha particle approaches a beryllium nucleus. When a beryllium nucleus with 4 protons and 5 neutrons absorbs an alpha particle with 2 protons and 2 neutrons the result is an unstable carbon nucleus with 6 protons and 7 neutrons. In the right frame the unstable carbon has emitted a neutron in the forward direction. What remains behind is a stable carbon with 6 protons and 6 neutrons.

Chadwick still imagined that the neutron he had identified was Rutherford's composite particle, a tightly bound system of an electron and a proton. But between 1932 and 1935 when he received the Nobel Prize in physics for "the discovery of the neutron" Chadwick changed his mind. After all, Heisenberg's newly discovered uncertainty principle made it impossible for an electron to be confined within the small volume of a neutron without having an unrealistically large energy. In addition, Chadwick had succeeded in measuring the mass of a neutron and found it to be larger than the sum of the masses of its presumed constituents, a proton and an electron – a result that doomed the idea of a composite neutron. The neutron must be an elementary particle.

Chadwick was well positioned to make his contribution, for he was a resourceful researcher trained in the art of cobbling together experiments from materials at hand. As a young man Chadwick was studying with Hans Geiger in Germany when World War I broke out. Geiger advised Chadwick, an Englishman, to leave Germany at once, but Chadwick delayed and was imprisoned along with other enemy aliens. In accordance with the Geneva Conventions the prisoners administered their own internal affairs. Thus, Chadwick lectured to his fellow inmates on radioactivity and conducted experiments using commercially available radioactive toothpaste as a source.

When World War II began Chadwick, now in England and a famous Nobel laureate, was asked to investigate the feasibility of building an atomic bomb based upon the fission of heavy nuclei. Subsequently he wrote the final draft of a report summarizing British efforts up to 1941. Eventually the British, exposed as they were to Luftwaffe bombing, abandoned the task of building their own bomb and offered their expertise to the Americans. Chadwick then became head of the British mission to the Manhattan Project and as such travelled to its various sites in America, moved his family to Los Alamos, and became a confidant of General Groves, the military engineer in charge of the Los Alamos effort. While Chadwick believed it was necessary for the allies to build an atomic bomb, he returned to Britain in 1948 disenchanted with the trend toward big, industrialized science of which the Manhattan Project was a prime example.

49. Nuclear Fission and Nuclear Fusion (1942)

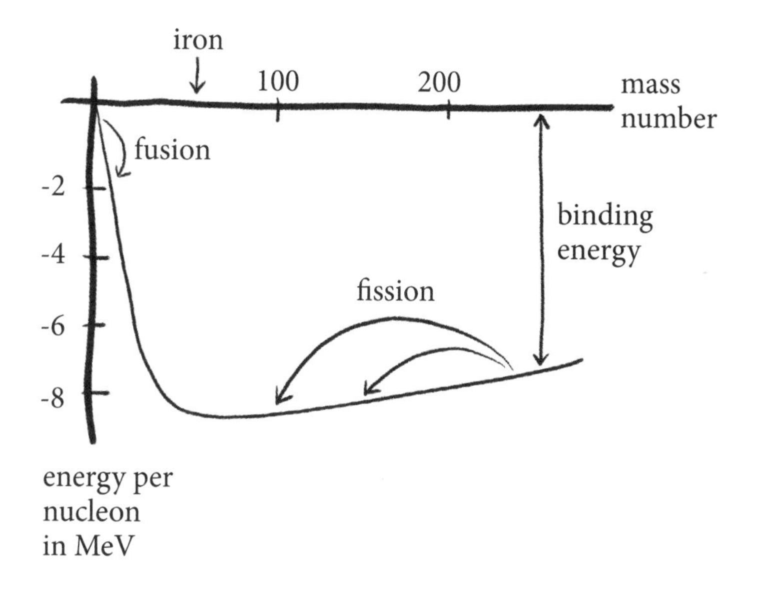

Figure 77

We are used to things happening in a certain way. Massive objects fall down – not up. Certain materials, like paper, burn easily. Others do not. We may be less familiar with nuclear fission and fusion, but the same general principle applies: a system (massive object, paper, nucleus) changes in a certain way (falls, burns, transforms) only when that system can lose energy in that change. This principle helps us understand nuclear fission and nuclear fusion. The first releases energy slowly in nuclear power plants and quickly in fission bombs. The second produces the energy beaming from our Sun and released explosively in a hydrogen bomb.

We know from Rutherford's gold foil experiment of 1912 that all of the positive charge and most of the mass of an atom is confined within a tiny nucleus several *fermis* across (1 fermi = 10^{-13} cm). By 1932 our current picture of the nucleus, as a spherical configuration of protons and neutrons, had emerged. Each proton is positively charged while each neutron is uncharged and slightly more massive than a proton. Thus, the number of protons within a nucleus is a measure of its charge, and the number of protons and neutrons in a nucleus, collectively called *nucleons*, is a measure of its mass. However, a question arises: given that like charges repel each other with a greater force the closer they are, what keeps the protons in a nucleus from flying apart? Evidently there is an even stronger attractive force that binds these nucleons together. Physicists call this force the *strong nuclear force*.

Competition between the repulsive electrostatic force among protons and the attractive strong nuclear force among nucleons determines the size, composition and stability of nuclei. The different natures of these competing forces make this competition interesting. The attractive force between nucleons is *strong* but *short range*. When two nucleons are within a couple of fermis of each other, they attract one another with a strong nuclear force that overwhelms any repulsive electrostatic force. But, when two nucleons are further apart than a couple of fermis, the attractive strong nuclear force between them vanishes. As a result, the strong nuclear force acts only between a nucleon and its nearest neighbours. In contrast, the repulsive electrostatic force between two protons diminishes slowly with separation – as the inverse of their separation squared – in the same way as does the gravitational force between the Sun and the Earth. The electrostatic, like the gravitational, force is said to be *long range*.

Now imagine adding more and more protons to an already large nucleus. As the number of protons grows, the net repulsive electrostatic force among all these protons grows, while the strong nuclear force keeping a nucleon (proton or neutron) bound to its nearest neighbours remains the same. For this reason, there is a natural limit to how many protons a nucleus can have. Uranium with 92 protons occupies that limiting position. Nuclei with more than 92 protons are unstable.

This competition also explains the stability properties of relatively light nuclei. Because these nuclei contain so few protons the net repulsive electrostatic force among them is weak relative to the strong nuclear force between neighbouring nucleons. Furthermore, the nucleons of hydrogen (one proton and no neutrons), of helium (two protons and two neutrons), and of lithium (three protons and four neutrons) are all on the surface of their relatively small, roughly spherical nuclei. Therefore, in these nuclei, each nucleon is not as strongly bound to its neighbours as it would be if completely surrounded by other nucleons. As more nucleons are added to a light nucleus, a typical nucleon gains more neighbours and, consequently, the whole system becomes more tightly bound. An iron nucleus, intermediate between lighter and heavier nuclei with 56 protons and neutrons, is the most stable.

This competition between repulsive electrostatic and attractive strong nuclear forces can also be thought of in terms of energy lost and gained. Think, for instance, of a rock rolling down a hill. The final configuration of the rock-Earth system has less energy than before. The energy "lost" is released in the kinetic energy of the rock as it falls. Ultimately this kinetic energy contributes to the thermal energy of the rock and the hillside.

Likewise the configuration of a system of nucleons after splitting apart (the fission process) or sticking together (the fusion process) has less energy than before. Because the number of nucleons after fission or fusion is the same as before, the energy per nucleon in the final configuration is less than in the energy per nucleon in the initial configuration. The nuclear energy lost is released in the kinetic energy of the fission or fusion products and in high-energy electromagnetic radiation.

The curve in figure 77, called the *curve of binding energy*, encapsulates this physics. Plotted is the energy per nucleon in various nuclei, from hydrogen to uranium, in the units typical of nuclear energy (MeV or million electron volts), versus the nuclear *mass number*: that is, the number of nucleons in the nucleus. The arrows indicate (on the left) the fusion of two, identical, light nuclei and (on the right) the fission of one heavy nucleus. Since the fusion or fission products have less energy per nucleon than before, energy is lost in the nuclear transformation and the final configurations are more tightly bound than before. The result of both fission and fusion is that nuclei with intermediate masses are more stable than lighter or heavier nuclei.

Nuclear fusions within the interior of the Sun are ultimately the source of all of our "non-nuclear" energy: fossil fuel, hydro, wind and solar power. But the fusion of two light nuclei does not happen as readily as the fission of a nucleus heavier than the stable form of uranium. After all, two light nuclei must overcome their electrostatic repulsion in order to fuse. Thus, fusion requires that light nuclei be driven together with a speed characteristic of the high temperature of the Sun's interior. Fission, on the other hand, occurs when a heavy nucleus – for instance, uranium with 235 nucleons or plutonium with 239 nucleons – absorbs an extra neutron. If among the fission fragments are several neutrons, one fission reaction can cause several others and each of those several more. The result: a self-sustaining chain reaction of nuclear fissions. Enrico Fermi and his team were the first, on December 2, 1942, to initiate and control a nuclear chain reaction of fissions.

The discovery and application of nuclear fission and nuclear fusion intertwine with the development of nuclear weapons. The story of their development tells of competition between the United States, Nazi Germany and Soviet Russia; sudden epiphanies while crossing a London street; a revelatory discussion between an aunt and her nephew, both physicists, on holiday together in Sweden; the contributions of Hungarian émigré scientists; letters to President Roosevelt signed by Albert Einstein; a nuclear reactor under the football stadium at the University of Chicago; British commandos destroying a (Nazi-controlled) Norwegian heavy water plant; a secret, multi-billion-dollar, nuclear industry in the United States; and Russian spies working incognito in the closed city of Los Alamos, New Mexico. Numerous books relate the human drama and the interesting science of this story. Two excellent ones are *The Making of the Atomic Bomb* by Richard Rhodes (1987) and *Nuclear Weapons: What You Need to Know* by Jeremy Bernstein (2008).

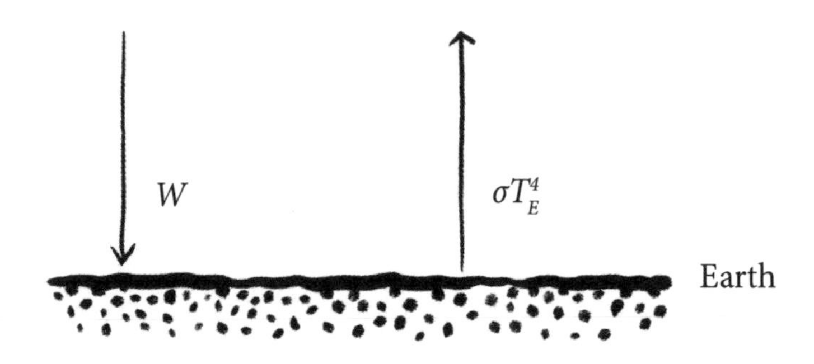

Figure 78

The Earth absorbs radiant energy from the Sun, transforms that energy into longer wavelength, infrared, thermal energy, and re-radiates this thermal energy skyward. By intercepting part of this re-radiated thermal energy and directing it back towards the Earth's surface, our atmosphere boosts the temperature of the Earth's surface above what it would be in its absence. Although such heating is commonly referred to as the *greenhouse effect*, actual greenhouses warm their contents in a different way – in particular, by inhibiting the circulation of air.

Consider two models: the *no-atmosphere* model (figure 78) and the *absorbing and radiating atmosphere* model (illustrated in figure 79). Together these illustrations show how our "global greenhouse" works. The Earth absorbs the energy of sunlight at an average rate of W watts per square metre. The no-atmosphere model assumes the Earth radiates into space as much energy as it receives from the Sun. Since all objects with temperature T radiate at σT^4 watts per square metre where σ is a universal constant, the Earth radiates at a rate $\sigma T_E^4[=W]$, where T_E is the average temperature of the Earth's surface. Thus T_E=($W/\sigma^{1/4}$), and so, given values of W and σ, T_E is found to be 254 degrees Kelvin – a frigid minus 19 degrees Celsius (minus 2 degrees Fahrenheit).

But the Earth does have an atmosphere, and that atmosphere absorbs thermal energy radiated from the Earth's surface and re-radiates that energy downward as well as upward at a rate of σT_E^4 watts per metre squared, where T_A is the average temperature of the atmosphere. Figure 79 shows the energy flows in the absorbing and radiating atmosphere model. Accordingly, the Earth and its atmosphere each radiates as much energy as each receives. Some algebra shows that, in this case, the average temperature of the Earth's surface is boosted over its *no-atmosphere* value by a factor of $2^{1/4}$[≈1.19]: that is, by approximately 19%. With this boost the average temperature of the Earth's surface T_E becomes 302 degrees Kelvin: that is, a warmish 29 degrees Celsius (84 degrees Fahrenheit). The current average temperature of the Earth's surface, 14.8 degrees Celsius (58.6 degrees Fahrenheit), lies between the temperatures implied by the no-atmosphere and the absorbing and radiating atmosphere models. Apparently, our atmosphere absorbs only part of the thermal energy radiated by the Earth's surface and is transparent to the remaining part.

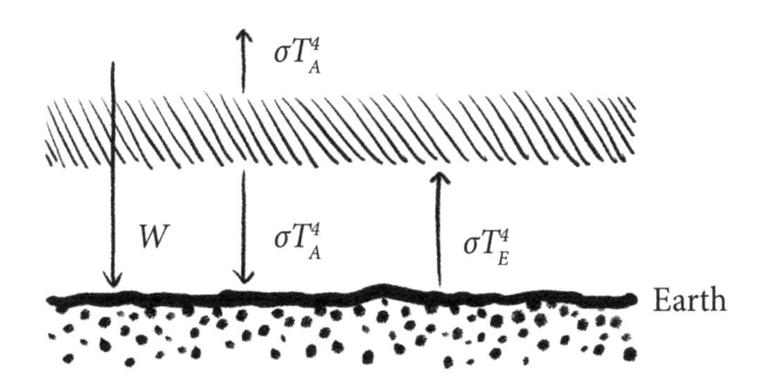

Figure 79

Of course, these models are simplifications that ignore changes within the atmosphere and do not account for other contributions to the temperature of the Earth's surface, such as the variable reflection of sunlight from clouds and snow. Still, they show that our atmosphere's ability to absorb and reradiate thermal energy is an important determinant of the temperature of the Earth's surface.

But what allows our atmosphere, a thin layer of gas containing less than one-millionth the mass of the Earth, to absorb and re-radiate this infrared radiation? The answer is, in large part, atmospheric carbon dioxide and water. While most of the atmosphere is nitrogen N_2 and oxygen O_2, argon is the third most numerous kind of molecule, and carbon dioxide CO_2 is the fourth. Water, H_2O, is also present in variable amounts, as well as smaller amounts of other gases. Notice that among these constituents CO_2 and H_2O molecules each contain three atoms. For instance, carbon dioxide CO_2 is composed of one carbon atom denoted C and two oxygen molecules denoted O_2. The more atoms in a molecule, the more ways its structure may flex and vibrate, and so resonate with and absorb the thermal radiation from objects with a temperature close to that of the Earth's surface.

While human beings do not directly control the amount of H_2O in the atmosphere, we do directly contribute to its CO_2 – chiefly by burning fossil fuels. In pre-industrial times the concentration of CO_2 in our atmosphere was 270 parts per million (ppm) or 0.0270%. Now our atmosphere has over 400 ppm CO_2, that is, over 0.0400%. By releasing more CO_2 into the atmosphere we increase its ability to absorb thermal radiation and, consequently, increase the average temperature of the Earth's surface.

The Swedish scientist Svante Arrhenius (1859–1927), who was the first in 1896 to note that increasing the number CO_2 of molecules in the atmosphere leads to global warming, considered such warming, in the main, a positive development that would prevent future ice ages and allow more of the Earth's surface to be cultivated. Ninety years later in 1988, James Hansen, then the director of the NASA Goddard Institute for Space Studies, warned of the hazards of global warming in testimony before committees of the United States Congress.

While the physics of the global greenhouse effect is simple, the phenomenon of global warming is not. For instance, as the temperature of the Earth's atmosphere increases, it absorbs more water vapour. This, of course, leads to even more heating, but more atmospheric H_2O also leads to more cloud cover, and since clouds reflect sunlight, clouds moderate, in some degree, this heating. Of course, we can do no experiments on the global system – or rather we can do only one irreversible experiment.

Since Hansen's testimony, climate scientists have collected much data and constructed complex numerical models that incorporate the physics relevant to climate change. They have checked their model predictions against those produced by independent teams of researchers and against historical data and quantified the uncertainty of those predictions. Their conclusion validates Hansen's warning more than it does Arrhenius's rosy prediction. Average surface temperatures are increasing at an alarming rate and human activity is a major cause of this increase.

But some resist this conclusion. The chair of the United States Senate Committee on Environment and Public Works in 2015, Senator James Inhofe of Oklahoma, quoting God's promise to Noah after the flood (Genesis 8:22 "While the Earth remains, seed time and harvest, cold and heat, summer and winter, day and night, shall not cease."), claims that it is arrogant for humans to believe they can change the Earth's climate. Indeed, arrogance may be an occupational hazard for scientists. After all, they invariably trust that natural processes can, in fact, be understood – a trust that sometimes devolves into arrogance.

Yet Senator Inhofe should know that God's will is often not done. We despoil rivers and lakes. We pollute land and air. We drive species into extinction. Given time we can do worse – or we can, by God's grace, do better. The late Reinhold Niebuhr (1892–1971), who was a wiser theologian than Senator Inhofe, urged us to pray for "the wisdom to distinguish between those things that can and those things that cannot be changed and for the courage to change those things that should be changed". Certainly, the distinction between those things that can and those things that cannot be changed is an important part of the understanding scientists seek. What we all need is the courage to change those things that should be changed.

51. Higgs Boson (2012)

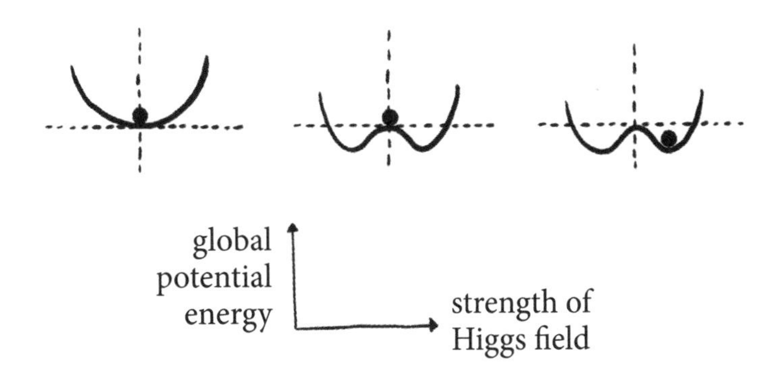

Figure 80

As we grow older some of us put on weight. We slowly begin to feel more massive, or, at least, heavier. And while feeling our weight or our mass is an everyday experience, we have probably never been moved to ask, "Why does anything have mass?" "Where does mass come from?"

Einstein's $E=mc^2$ or, equivalently, $m=E/c^2$ provides one kind of answer. Evidently, anything with energy E has mass m in the amount E/c^2. But consider an elementary particle, such as an electron, isolated and at rest: that is, a particle with no parts, no apparent spatial extent, no obvious energy and no motion. Yet an electron acts as if it has a tiny *rest mass* of $9 \cdot 10^{-28}$ grams. Our question then becomes, "Why do elementary particles have rest mass?" Or, if you prefer, "Why do isolated elementary particles at rest have energy?"

Before 1930 scientists knew of only two elementary particles: the electron and the proton. Then James Chadwick discovered the neutron and shortly thereafter Wolfgang Pauli reasoned that another kind of elementary particle, the neutrino, must exist. Today we know that protons and neutrons are themselves each composed of three *quarks*. According to the *standard model* of particle physics, quarks are elementary particles that, like the electron, carry electric charge, have mass, and have no spatial extent. Electrons and neutrinos are also elementary particles but of a class, distinct from quarks, called *leptons*. *Photons* belong to yet a third class of elementary particles.

The *theoretical* concept of mass has emerged as a product of our long-standing effort to unify the fundamental forces that act among the elementary particles. Just as in the mid-19th century James Clerk Maxwell, building upon Michael Faraday's ideas, unified the separate theories of electricity and magnetism into a single theory of electromagnetism, so also, in the early 1960s, were the separate theories of electromagnetism and the weak nuclear force unified into a single *electro-weak* theory.

The main building blocks of the electro-weak theory are the several symmetries assigned to these forces. A *symmetry* is a property that is preserved while something else changes. For instance, we believe that the basic laws of physics are the same in every part of the universe. Or, as a physicist would say, these laws are symmetrical under translation from one part of the universe to another. Unfortunately, the simplest version of the electro-weak theory with its embedded symmetries requires that all elementary particles have zero rest mass – a claim we know to be false. For, while some elementary particles, like the photon, are massless in this sense, others, like the electron, are not.

How then can the simplest description of the electro-weak theory be modified so that some elementary particles have mass? One answer is that the universe contains a *Higgs field* that allows those particles with which it interacts to acquire mass.

Such acquisition is sometimes compared to a celebrity trying to move through a crowd while stopping frequently to shake hands, receive compliments, and pose for photos. Because the celebrity (the particle) interacts strongly with the crowd (the Higgs field), he or she acquires mass – at least of a certain kind. In contrast, an unknown person hardly interacts with the crowd at all just as if he or she has none of this "celebrity mass". Quarks and leptons similarly interact, in varying degrees, with the Higgs field and, consequently, acquire mass in various amounts. On the other hand, because a photon does not interact with the Higgs field, it has no rest mass and travels at the speed of light.

The Higgs field and the mechanism with which it confers mass on otherwise massless particles makes the electro-weak theory consistent with known particle rest masses and also with our everyday experience of mass. In 2012, the Higgs boson, itself a characteristic excitation of the Higgs field, was observed in the detectors of the Large Hadron Collider of the CERN (after the French *Conseil Européen pour la Recherche Nucléaire*) laboratory in Geneva, Switzerland.

Apparently, early in its history, the universe transitioned from a state with no Higgs field to one with a Higgs field – a transition initiated by the emergence of a "Mexican hat" global potential energy, shown above in the middle and rightmost panels of figure 80. The horizontal position of the black circle indicates the strength of the universe's Higgs field. The black circle initially resides in the centre of the (leftmost) diagram indicating no Higgs field. When the Mexican hat global potential emerged (centre diagram), the universe was for a short time in an unstable state. Then the universe transitioned to a stable state with non-vanishing Higgs field (rightmost diagram). These diagrams are meant to suggest the rolling of a marble, initially perched on the crest of the Mexican hat potential, into the trough between the hat's crest and its upturned brim. In this way the fundamental forces, as represented by the Mexican hat potential, retain their symmetries while the state of the universe, represented by the position of the black circle, "breaks" this symmetry.

Peter Higgs (1929–) and François Englert (1932–) shared the 2013 Nobel Prize in physics for, in 1964, postulating the Higgs field, describing the Higgs mechanism, and predicting the Higgs boson. That Higgs's name alone is attached, in proprietary fashion, to these discoveries may well embarrass Peter Higgs. For, while Higgs certainly deserves a Nobel Prize, he is only one of several theoreticians who independently and nearly simultaneously came to the same conclusions.

Afterword

The principle with which I have selected concepts to explore in these essays, *those that can be seen*, favours older explanations such as of lunar phases and of submerged bodies over more recent, less visualizable ones such as of the global greenhouse effect and of the Higgs boson. Other important contributions to 20th- and 21st -century physics – probability distributions, black holes, chaos, entanglement and gravity waves – did not make the cut because I could not easily represent them in a drawing. No doubt, our drive toward inventing new ways to see physics will, in time, allow these topics to be grasped visually.

Because *Seeing Physics* is composed of 51 separate essays, the connections among them may not be apparent. Consider, for instance, that the theorists who predicted the Higgs boson (essay 51) built upon an approach pioneered by Enrico Fermi who, in the course of modelling beta decay (essay 47), fashioned the first version of a field theory of the elementary particles and their fundamental forces. Fermi's field theory, in turn, owes much to Maxwell's 1865 theory of electromagnetism (essay 37) and to Einstein's 1905 theory of special relativity. Yet Maxwell's electromagnetism incorporates Faraday's concept of electric and magnetic field lines (essay 36) while Einstein's theory of special relativity generalizes the relativity of Galileo. The kinematics which embeds Galilean relativity (essays 20 and 21) appeals to observations on falling bodies made 50 years earlier by Simon Stevin (essay 15) and 1,000 years earlier by John Philoponus (essay 10) and, also, to the geometrical language invented by Oresme and the Merton College scholars (essay 12). Philoponus's observations critique Aristotle's theory of motion (essay 5) just as Aristotle had earlier critiqued Thales' cosmology (essay 1). Other threads of understanding, each contribution depending upon others, could be traced through the essays.

Only physics and astronomy, among the empirical sciences, can claim an intellectual tradition of such compass: 2,600 years from Thales to Higgs. And no other science appeals so directly to the visible world. The visual emblems of this tradition and their human and historical context define the content of *Seeing Physics*.

Notes and Sources

Antiquity: Essays 1–9

3.4 "The sun puts the shine in the Moon" Nahm, *Selections from Early Greek Philosophy* (1964), 143.

4.2 "You cannot step twice into the same river . . . " Nahm, *Selections from Early Greek Philosophy* (1964), 70.

4.2 ". . . as when a young girl, playing with a clepsydra. . ." Curd, *A Presocratics Reader* (2011), 97.

5.3 "One of the greatest philosophers and scientists of all times." Sarton, *Introduction to the History of Science* (1927), 127.

5.3 "carried on immense botanical, zoological, and anatomical investigations . . ." Sarton, *Introduction to the History of Science* (1927), 127.

7.1 " a line is breadthless length", "A straight line is a line that lies evenly with the points on itself." Euclid, *The Elements* (1956), 153.

7.1 "Things which are equal to the same thing are also equal to one other." Euclid, *The Elements* (1956), 155.

7.2 ". . . let geese, Gabble and hiss, but heroes seek release . . ." From "Euclid Alone Has Looked on Beauty Bare" by Edna St Vincent Millay. Salter and Stallworth, *Norton Anthology of Poetry* (2005), 1383.

Middle Ages: Essays 10–13

10.2 ". . . this view of Aristotle's is completely erroneous . . ." Lindberg, *The Beginnings of Western Science* (1992), 305.

Early Modern Period: Essays 14–31

14.3 "I finally discovered . . . that if the movements . . ." Copernicus, *On the Revolutions of the Heavenly Spheres* (1952), 508.

14.4 ". . . machinery of the world," "Most Orderly Workman of all." Copernicus, *On the Revolutions of the Heavenly Spheres* (1952), 508.

15.2 "If you can get an epitaph like that . . ." Feynman, *The Feynman Lectures on Physics* (1963), vol. I, 4–5.

18.2 "There remains the matter which in my opinion . . ." Drake, *Discoveries and Opinions of Galileo* (1957), 50–51.

18.3 ". . . that the revolutions are swifter in those planets . . ." Drake, *Discoveries and Opinions of Galileo* (1957), 57.

18.3 "Here we have a fine and elegant argument . . ." Drake, *Discoveries and Opinions of Galileo* (1957), 57.

19.3 "Thee, O Lord Creator . . ." Sobel, *A More Perfect Heaven* (2011), 211.

19.3 "If I have been drawn into rashness . . ." Sobel, *A More Perfect Heaven* (2011), 211.

20.1 ". . . hard to say whether the qualities. . . ." Drake, *Discoveries and Opinions of Galileo* (1957), 5.

20.2 "I am still at sea, he says,. . ." Galileo, *Two New Sciences* (1952), 158.

20.2 "It will not be beyond you. . ." Galileo, *Two New Sciences* (1952), 158.

20.3 "Aristotle says, "An iron ball of one hundred pounds . . ."" Galileo, *Two New Sciences* (1952), 158.

21.1 ". . . relinquish altogether the said opinion . . ." and ". . . to hold teach or defend it in any way. . . ." Crombie, *Medieval and Early Modern Science* (1959), vol. II, 212–213.

22.3 "Thus a small dog could probably carry . . ." Galileo, *Two New Sciences* (1952), 187.

23.2 ". . . the defect was not in the pump. . ." Galileo, *Two New*

Sciences (1952), 137–138.

23.2 "We live submerged at the bottom of an ocean of . . ." Evangelista Torricelli in a letter to Michelangelo Ricci as excerpted in Boynton, *The Beginnings of Modern Science* (1948), 227.

23.4 "Does nature abhor a vacuum more . . ." Blaise Pascal in *The Great Experiment on the Weight of the Mass of the Air* in Boynton, *The Beginnings of Modern Science* (1948), 231–241.

24.3 The table of values is taken from Boyle's *A Defence of the Doctrine Touching the Spring and Weight of Air* as excerpted in Boynton, *The Beginnings of Modern Science* (1948), 246.

24.5 ". . . to the injuries of both parties, and the protection of neither." Wojcik, *Robert Boyle and the Limits of Reason* (1997), 13.

24.5 ". . . a very great caution. . . ." Wojcik, *Robert Boyle and the Limits of Reason* (1997), 13.

25.2 For Westfall's belief that the young Newton had a powerful patron see Westfall, *Never at Rest* (1980), 102.

25.3 Newton describes his *experimentium crucis* in *A New Theory of Light and Colors* as excerpted in Boynton, *The Beginnings of Modern Science* (1948), 148–156.

25.3 For Westfall's confession that the more he knew of Newton the more alien the latter became, see Westfall, *Never at Rest* (1980), ix.

29.2 ". . . the true philosophy, in which one conceives . . ." Huygens in *Treatise on Light* (1952), 554.

29.3 Vasco Ronchi argues that Huygens did not consider diffraction as evidence for his view of the nature of light because he could not quantify the contribution of the secondary waves to the intensity of the envelope. See Ronchi, *The Nature of Light* (1970), 202.

Nineteenth Century: Essays 32–37

32.3 "... the accuracy, with which the general law of interference ... " Young, *A Course of Lectures on Natural Philosophy* (1845), Lecture XXXIX, 370.

32.4 "... a man alike eminent in almost every department ..." "Sketch of Dr. Thomas Young", *Popular Science Monthly* (1874), 360.

33.1 "And when the rain has wet the kite and twine ..." Boynton, *The Beginnings of Modern Science* (1948), 320.

33.3 "From the preceding facts we may likewise collect ..." Dibner, *Oersted and the Discovery of Electromagnetism* (1962), 75.

33.4 "My heart leaps up when I behold ..." From "My Heart Leaps Up" by William Wordsworth. Salter and Stallworth, *Norton Anthology of Poetry* (2005), 796.

34.2 "To take away from England ..." Carnot, *Reflections on the Motive Power of Fire* (2005), 4.

34.3 "The production of motion in steam engines. ..." Carnot, *Reflections on the Motive Power of Fire* (2005), 6.

35.3 "It would be difficult to describe the surprise and astonishment. ..." Boynton, *The Beginnings of Modern Science* (1948), 195.

35.3 "It is hardly necessary to add ..." Boynton, *The Beginnings of Modern Science* (1948), 198.

36.1 "... trembled and grew cold .... There had to be something behind objects that lay deeply hidden." Pais, *Subtle is the Lord* (1982), 37.

36.2 "The Physical Character of the Lines of Magnetic Force (1852)." Fisher, *Faraday's Experimental Researches in Electricity* (2001), 563–599.

36.3 "speculations" and "strict line of reasoning." Fisher, *Faraday's Experimental Researches in Electricity* (2001), 563.

Twentieth Century and Beyond: Essays 38–51

38.2 "There was once a sailor on a vessel in New York harbor . . ." Stuewer, *The Compton Effect* (1975), 43.

38.3 ". . . the theory of light would be thrown back by centuries." Holton and Brush, *Physics: The Human Adventure* (2002), 401.

38.4 ". . . especially for his discovery of the law of the photoelectric effect." *Nobel Citations* (1922).

38.4 "All these 50 years of pondering have not brought me any closer to answering the question, 'What are light quanta?' " Pais, *Subtle is the Lord* (1982), 382.

39.1 Robert Brown, who thoroughly investigated Brownian motion, is usually credited with its discovery in 1827. However, Jan Ingenhousz quite clearly described the Brownian motion of coal dust particles on the surface of alcohol in 1785.

39.3 "I at any rate am convinced that He . . ." Einstein and Born, *The Born-Einstein Letters* (2004), xxii.

39.3 "He has seen more clearly than anyone before him . . ." Schlipp, *Albert Einstein; Philosopher-Scientist* (1949), 163–164.

39.4 ". . . based on different experiences in our work and life . . ." Schlipp, *Albert Einstein; Philosopher-Scientist* (1949), 177.

40.2 "I was brought up to look at the atom as a nice hard fellow . . ." Keller, *The Infancy of Atomic Physics* (2006), 9.

40.3 "It was quite the most incredible event that has ever happened to me . . ." Pais, *Inward Bound* (1986), 189.

41.2 "I could not recognize my own work in the reports." Pais, *Inward Bound* (1986), 39.

41.2 ". . . every Professor in Europe is now on the warpath." Pais, *Inward Bound* (1986), 39.

41.4 "suddenly . . . perceived the way which subsequently proved to be the shortest path to success." *Nobel Citations* (1914).

41.4 " . . . his discovery of the diffraction of X-rays by crystals." *Nobel Citations* (1914).

42.3 "While it is too early to say whether the theories of Bohr . . ." Pais, *Niels Bohr's Times* (1991), 153.

43.1 "Doomed youth" and "as cattle" from "Anthem for Doomed Youth" by Wilfred Owen. Salter and Stallworth *Norton Anthology of Poetry* (2005), 1386.

43.3 "Overthrown" and "knocked out." Pais, *Subtle is the Lord* (1982), 306–309.

44.2 "In sum, one can say that there is hardly one among the great problems . . ." Pais, *Subtle is the Lord* (1982), 382.

44.3 ". . . quite likely never read Einstein's 1905 paper . . ." Steuwer, *The Compton Effect* (1975), 217–218.

45.3 "He has lifted a corner of the great veil." Moore, *Schrödinger, Life and Thought* (1989), 187.

45.4 ". . . for his discovery of the wave nature of electrons." *Nobel Citations* (1929).

46.3 "Here is the letter that has destroyed my universe." Payne-Gaposchkin, *An Autobiography and Other Recollections* (1997), 209.

47.2 "This theory [of transformation] is found to account . . ." Pais, *Inward Bound* (1986), 113.

47.2 "The underlying physical laws necessary for the mathematical theory . . ." Dirac, "Quantum Mechanics of Many-Electron Systems", 714.

47.3 "I have done something very bad today . . ." Solomey, *The Elusive Neutrino* (1997), 14.

47.4 "We are happy to inform you that we have definitely detected neutrinos." Solomey, *The Elusive Neutrino* (1997), 65.

48.3 "The results, and others I have obtained . . ." Schweber, *Nuclear Forces* (2012), 220.

48.3 ". . . for the discovery of the neutron." *Nobel Citations* (1935).

50.4 For an example of Senator James Inhofe's rhetoric see the Nov. 12, 2014 issue of the *New York Times*.

50.4 "While the earth remains . . ." Genesis 8:22, Revised Standard Version.

50.5 "The wisdom to distinguish between those things . . ." A paraphrase of the widely reproduced "serenity prayer" typically ascribed to Reinhold Niebuhr.

Bibliography

Archimedes. *The Works of Archimedes*. Edited and translated by T. L. Heath. New York: Dover Publications, no date.

Arrhenius, Svante. "On the Influence of Carbonic Acid in the Air Upon the Temperature of the Ground", *Philosophical Magazine and Journal of Science*, 5th ser. (April 1896): 237–276.

Bernstein, Jeremy. *Nuclear Weapons: What You Need to Know.* Cambridge: Cambridge University Press, 2007.

Boynton, Holmes, ed. *The Beginnings of Modern Science*. Roslyn, NY: Walter J. Black, Inc., 1948.

Carnot, Sadi. *Reflections on the Motive Power of Fire and other papers on the Second Law of Thermodynamics.* New York: Dover Publications, 2005.

Copernicus, Nicolaus. *On the Revolutions of the Heavenly Spheres.* Translated by Charles Glenn Wallis. Chicago, IL: Encyclopedia Britannica, 1952.

Crombie, A. C. *Medieval and Early Modern Science*, 2 vols. Garden City, NY: Doubleday Anchor, 1959.

Curd, Patricia, ed. *A Presocratics Reader: Selected Fragments and Testimonia.* 2nd ed. Translated by Richard D. McKirahan. Indianapolis, IN: Hacket Publishing, 2011.

Dibner, Bern. *Oersted and the Discovery of Electromagnetism.* New York: Blaisdell Publishing Company, 1962.

Dirac, P. A. M. "Quantum Mechanics of Many-Electron Systems", *Proceedings of the Royal Society*, A123 (1929): 714.

Drake, Stillman. *Discoveries and Opinions of Galileo.* New York: Doubleday, 1957.

Dreyer, J. L. E. *A History of Astronomy from Thales to Kepler,* chaps. XIV-XV. New York: Dover Publications, 1953.

Einstein, Albert. *Albert Einstein: Investigations on the Theory of the*

Brownian Movement. Edited by R. Furth. Translated by A. D. Cowper. New York: Dover Publications, 1956.

Einstein, Albert. "Concerning an Heuristic Point of View Toward the Emission and Transformation of Light", *American Journal of Physics*, vol. 33, no. 5 (May 1965): 1–16. Originally published in *Annalen de Physik* 17 (1905): 132–148.

Einstein, Albert and Max Born. *The Born-Einstein Letters 1916–1915: Friendship, Politics, and Physics in Uncertain Times.* New York: Palgrave McMillan, 2004.

Euclid. *The Elements.* Edited and translated by T. L. Heath. New York: Dover Publications, 1956.

Feynman, Richard. *The Feynman Lectures on Physics*, 3 vols. Reading, MA: Addison-Wesley, 1963.

Fisher, Howard J. *Faraday's Experimental Researches in Electricity: Guide to a First Reading.* Santa Fe, NM: Green Lion Press, 2001.

Fisher, Irene. "Another Look at Eratosthenes' and Posidonius's Determination of the Earth's Circumference", *Quarterly Journal of the Royal Astronomical Society* 16 (1975): 152–167.

Galilei, Galileo. *Two New Sciences.* Translated by Henry Crew and Alfonso de Salvio. Chicago, IL: Encyclopedia Britannica, 1952.

Gillespie, Charles Coulston, ed. *Complete Dictionary of Scientific Biography.* Detroit, MI: Charles Scribner's Sons, 2007.

Gleick, James. *Isaac Newton.* New York: Pantheon Books, 2003.

Grant, Michael. *Physical Science in the Middle Ages.* Cambridge: Cambridge University Press, 1977.

Gullen, Michael. *Five Equations that Changed the World.* New York: Hyperion, 1995.

Haskins, Charles Homer. *The Rise of the Universities.* Ithaca, NY: Cornell University Press, 1957.

Heath, T. L. *A History of Greek Mathematics.* New York: Dover Publications, 1981.

Holton, Gerald and Brush, Stephen. *Physics: The Human Adventure: From Copernicus to Einstein.* NJ: Rutgers University

Press, 2001.

Huygens, Christiaan. *Treatise on Light.* Translated by Sylvanus P. Thompson. Chicago, IL: Encyclopedia Britannica, 1952.

James, Ioan. *Remarkable Physicists.* Cambridge: Cambridge University Press, 2004.

Keller, Alex. *The Infancy of Atomic Physics.* Mineola, NY: Dover Publications, 2006.

Lemons, Don S. *Mere Thermodynamics.* Baltimore, MD: Johns Hopkins University Press, 2008.

Lindberg, David C. *The Beginnings of Western Science.* Chicago, IL: Chicago University, 1992.

Magie, William Francis. *A Source Book in Physics.* New York: McGraw-Hill Book Company, 1935.

Mahon, Basil. *The Man Who Changed Everything.* Chichester, UK: Wiley, 2003.

Maxwell, James Clerk. *A Treatise on Electricity and Magnetism*, 2 vols. New York: Dover Publications, 1954.

Moore, Walter. *Schrödinger, Life and Thought.* Cambridge: Cambridge University Press, 1989.

Munitz, Milton K., ed. *Theories of the Universe.* New York: Macmillan, 1957.

Nahm, Milton. *Selections from Early Greek Philosophy.* 4th ed. New York: Meredith Publishing Company, 1964.

Nobel Citations. Various years, http://www.nobelprize.org/nobel_prizes/physics/laureates/ (accessed July 27, 2016).

Pais, Abraham. *Subtle is the Lord: The Science and the Life of Albert Einstein.* New York: Oxford University Press, 1982.

Pais, Abraham. *Inward Bound: Of Matter and Forces in the Physical World.* New York: Oxford University Press, 1986.

Pais, Abraham. *Niels Bohr's Times in Physics, Philosophy, and Polity.* Oxford: Oxford University Press, 1991.

Payne-Gaposchkin, Cecilia. *An Autobiography and Other Recollections.* Cambridge: Cambridge University Press, 1997.

Peacock, George. *Life of Thomas Young.* London: John Murray,

1855.
Perrin, Jean. *Atoms.* Translated by D. L. Hammick. New York: D. Van Nostrand, 1916.
Randall, Lisa. *Higgs Discovery: The Power of Empty Space.* New York: Harper Collins, 2013.
Rhodes, Richard. *The Making of the Atomic Bomb.* New York: Simon and Schuster, 1987.
Ronchi, Vasco. *The Nature of Light.* Translated by V. Barocas. Cambridge, MA: Harvard University Press, 1970.
Salter, M. J., M. Ferguson, and J. Stallworth, eds. *Norton Anthology of Poetry*. 5th ed. New York: W. W. Norton & Company, 2005.
Sarton, George. *Introduction to the History of Science,* Vol. I, *From Homer to Omar Khayyam.* Washington, DC: Williams and Wilkens for the Carnegie Institution of Washington, 1927.
Schiffer, Michael. *Draw the Lightning Down: Benjamin Franklin and the Electrical Technology of the Age of Enlightenment.* Berkeley, CA: University of California Press, 2003.
Schlipp, Paul Arthur, ed. *Albert Einstein; Philosopher-Scientist.* La Salle, IL: Open Court, 1949.
Schweber, Silvan S. *Nuclear Forces: The Making of the Physicist Hans Bethe.* Cambridge, MA: Harvard University Press, 2012.
Simonyi, Károly. *A Cultural History of Physics.* Translated by David Kramer. New York: CRC Press, 2012.
Simpson, Thomas K. *Maxwell on the Electromagnetic Field: A Guided Study.* New Brunswick, NJ: Rutgers University Press, 1998.
"Sketch of Dr. Thomas Young." *Popular Science Monthly* 5 (July 1874): 360.
Sobel, Dava. *A More Perfect Heaven: How Copernicus Revolutionized the Cosmos.* New York: Walker and Company, 2011.
Solomey, Nicholas. *The Elusive Neutrino: A Subatomic Detective Story.* New York: W. H. Freeman and Company, 1997.
Stevin, Simon. *The Principal Works of Simon Stevin,* vol. 1. Ed. E. J.

Dijksterhuis. Amsterdam: C. V. Svets and Zeitlinger, 1955.
Stuewer, Roger H. *The Compton Effect.* New York: Neale Watson Academic Publications, 1975.
Taylor, Lloyd W. *Physics, The Pioneer Science*, 2 vols. New York, NY: Dover Publications, 1941.
Westfall, Richard S. *Never at Rest. A Biography of Isaac Newton.* Cambridge: Cambridge University Press, 1980.
Wheelwright, Philip. *The Presocratics.* Englewood Cliffs, NY: Macmillan, 1966.
Wojcik, Jan W. *Robert Boyle and the Limits of Reason.* Cambridge: Cambridge University Press, 1997.
Young, Thomas. *A Course of Lectures on Natural Philosophy and the Mechanical Arts.* London: Taylor and Walton, 1845.

Index

(page numbers in italics refer to diagrams)

N

O

Q

R